Biological Diversity and International Law

Mar Campins Eritja • Teresa Fajardo del Castillo
Editors

Biological Diversity and International Law

Challenges for the Post 2020 Scenario

Springer

Editors
Mar Campins Eritja
Universitat de Barcelona, Faculty of Law
Barcelona, Spain

Teresa Fajardo del Castillo
Universidad de Granada, Faculty of Law
Granada, Spain

ISBN 978-3-030-72960-8 ISBN 978-3-030-72961-5 (eBook)
https://doi.org/10.1007/978-3-030-72961-5

This Springer imprint is published by the registered company Springer Nature Switzerland AG.
The registered company address is: Gewerbestrasse 11, 6330 Cham, Switzerland

Acknowledgements

This book is an outcome of the research project "Biological Diversity and International Law: Emerging Issues and Trends, Interaction of Legal Regimes and Key Policy Challenges", (*Diversidad Biológica y Derecho Internacional: Nuevos Problemas y Tendencias, Interacción de Regímenes Jurídicos y Principales Retos Normativos*) (DER2017-85406-P), funded by the Spanish Government.

Contents

1 Introduction 1
Mar Campins Eritja and Teresa Fajardo del Castillo

Part I Principles Inspiring Biodiversity-Related Conventional Law and Governance

2 Principles and Approaches in the Convention on Biological Diversity and Other Biodiversity-Related Conventions in the Post-2020 Scenario 15
Teresa Fajardo del Castillo

3 Gaps in International Biodiversity Law and Possible Ways Forward 35
José Juste Ruiz

4 Opportunities and Challenges for Synergies Across Biodiversity-Related Conventions in the Context of Human Health and Zoonotic Diseases: The Role of Scientific Advisory Bodies 57
Royal C. Gardner

Part II Biodiversity Interactions with Global Conventional Regimes

5 Conservation and Sustainable Use of Biodiversity in the International Regulation of Trade in Goods 79
Xavier Fernández-Pons

6 Is Everything Marketable in International Trade? Public Health Issues in International Trade of Wildlife 101
Josep Maria de Dios Marcer

7 Access to and Benefit-Sharing of Genetic Resources: Rationale, Review and Prospects 119
Xavier Pons Rafols

Part III Some Challenges for Sustainability and Marine Biodiversity

8 Sub-National Island Jurisdictions and Biodiversity Protection: Moving Towards Greater Resilience and Sustainability 141
Francesco Sindico

9 The Arctic Ocean: Ecosystem Approach in a Context of Extreme Vulnerability 157
Mar Campins Eritja

10 Governance of the Caribbean Marine Biodiversity: Complex Challenges for a Complex Region 179
Luis E. Rodríguez-Rivera

11 The Convention on Biological Diversity and the Mediterranean Sea: A Beacon for International Environmental Governance in the Region? 205
Guillaume Futhazar

Editors and Contributors

About the Editors

Mar Campins Eritja is Full Professor of Public International Law at the Universitat de Barcelona. Her main areas of teaching and research are EU Law and European and International Environmental Law. She has lead and participated in various competitive research projects and has published extensively in these areas. She is the Holder of Jean Monnet Chair on EU Environmental Law at the University of Barcelona. She is Fulbright Scholar at American University Washington College of Law and Visiting Professor at the Universities of Ottawa, Dalhousie, Montreal, Nova Southeastern, Puerto Rico and Rovaniemi. She is member of the Executive Board of the UB Research Water Institute (IdRA), the Centre for Environmental Law Studies of Tarragona (CEDAT) at the University Rovira i Virgili and the Spanish Observatory of Environmental Policies (OPA). She is member of the Editorial Board of the *Revista Catalana de Dret Ambiental* since 2009 and of the *Revista Catalana de Dret Public* (2012–2016).

Teresa Fajardo del Castillo is Associate Professor at the Universidad de Granada where she teaches Public International Law at the Faculty of Law and International and European Environmental Law at the Science Faculty. She is an associated member of Jean Monnet Chair on EU Environmental Law at the University of Barcelona. Her research interests lie in the field of International and European Environmental Law, EU migration law and policy and Soft Law. She has been a Visiting Researcher at the Universities of Brussels, Geneva, King's College of London and the Cambridge Centre for Environment, Energy and Natural Resource Governance (C-ENRG). She has drawn up reports for the European Commission and the European Parliament on Wildlife Crime and Environmental Crime.

Contributors

Mar Campins Eritja Universitat de Barcelona, Faculty of Law, Barcelona, Spain

Teresa Fajardo del Castillo Universidad de Granada, Faculty of Law, Granada, Spain

Jose Juste Ruiz Universitat de Valencia, Faculty of Law, Valencia, Spain

Royal C. Gardner Stetson University, College of Law, Gulfport, FL, USA

Xavier Fernández-Pons Universitat de Barcelona, Faculty of Law, Barcelona, Spain

Josep Maria de Dios Marcer Universitat Autònoma de Barcelona, Faculty of Law, Bellaterra, Spain

Xavier Pons Rafols Universitat de Barcelona, Faculty of Law, Barcelona, Spain

Francesco Sindico University of Strathclyde, Law School, Glasgow, UK

Luis E. Rodriguez-Rivera Universidad de Puerto Rico, Law School, San Juan, PR, USA

Guillaume Futhazar Max Planck Institut für ausländisches öffentliches Recht und Völkerrecht (Max Planck Institute for Comparative Public Law and International Law), Heidelberg, Germany

List of Abbreviations

AB	Appellate Body
ABNJ	Areas Beyond National Jurisdiction
ABS	Access and Benefit Sharing
ACAP	Arctic Contaminants Action Program
ACCOBAMS	Agreement on the Conservation of Cetaceans of the Black Sea, Mediterranean Sea and Contiguous Atlantic Area
ACS	Association of Caribbean States
AEWA	Agreement on the Conservation of African-Eurasian Migratory Waterbirds
AHTEG	Ad Hoc Technical Expert Group
AMAP	Arctic Monitoring and Assessment Programme
AMSA	Arctic Marine Shipping Assessment
BBNJ	Biological Diversity of Areas Beyond National Jurisdiction
BLG	Liaison Group of Biodiversity-related Conventions
BSF	Benefit-sharing fund
CAFF	Conservation of Arctic Flora and Fauna Working Group
CARICOM	Caribbean Community
CBD	Convention on Biological Diversity
CBMP	Circumpolar Biodiversity Monitoring Programme
CCCFP	Agreement establishing the Caribbean Community Common Fisheries Policy
CDC	Centers for Disease Control and Prevention
CEPF	Critical Ecosystem Partnership Fund
CGRFA	FAO Commission on Genetic Resources for Food and Agriculture
CHM	Common heritage of mankind
CIBH	Caribbean Islands Biodiversity Hotspot
CITES	Convention on International Trade in Endangered Species of Wild Fauna and Flora
CLME	Caribbean Large Marine Ecosystem

CLME+	Caribbean Large Marine Ecosystem and Adjacent Areas
CMS	Convention on the Conservation of Migratory Species of Wild Animals
CoP	Conference of Parties
CoP-MoP	Conference of the Parties serving as the Meeting of the Parties
COVID-19	Coronavirus Disease
CPAN	Circumpolar Protected Areas Network
CPTPP	Comprehensive and Progressive Agreement for Trans-Pacific Partnership
CSAB	Chairs of the Scientific Advisory Bodies
CRFM	Caribbean Regional Fisheries Mechanism
CTE	Committee on Trade and Environment
DNA	Deoxyribonucleic acid
DSI	Digital sequence information
DSI/GSD	Digital sequence information/genetic sequence data
EBSA	Ecologically or Biologically Significant Marine Areas
EC	European Community
ECJ	European Court of Justice
EMIT	Environmental Measures and International Trade
EPPR	Emergency Prevention, Preparedness and Response Working Group
ETP	Eastern Tropical Pacific Ocean
EU	European Union
EUCJ	European Union Court of Justice
FAO	Food and Agriculture Organization
FCNMPA	Framework for a Circumpolar Network of Marine Protected Areas
FPIC	Free, prior, informed Consent
GATT	General Agreement on Tariffs and Trade
GCR	Greater Caribbean Region
GEF	Global Environment Facility
GFCM	General Fisheries Council for the Mediterranean
GMOs	Genetically modified organisms
GOM LME	Gulf of Mexico Large Marine Ecosystems
GPA	Government Procurement Agreement
GSD	Genetic sequence data
HLPAI	Highly pathogenic avian influenza
ICJ	International Court of Justice
ILA	International Law Association
ILC	International Law Commission
ILCs	Indigenous and local communities
ILO	International Labour Organization
ILRI	International Livestock Research Institute

ILUC	Indirect land-use change
IMO	International Maritime Organization
IPBES	Intergovernmental Science-Policy Platform on Biodiversity and Ecosystem Services
IPPC	International Plant Protection Convention
IPCC	Intergovernmental Panel on Climate Change
ITLOS	International Tribunal for the Law of the Sea
ITPGRFA	International Treaty on Plant Genetic Resources for Food and Agriculture
ISA	International Seabed Authority
IUCN	International Union for the Conservation of Nature
IWC	International Whaling Convention
JLG	Joint Liaison Group between the three Rio conventions (CBD, UNFCC and Desertification)
LMEs	Large marine ecosystems
LMEH	Large marine ecosystems hub
LPAI	Low pathogenic avian influenza
MAP	Mediterranean action plan
MARPOL	International Convention for the Prevention of Pollution from Ships
MAT	Mutually Agreed Terms
MEAs	Multilateral environmental agreements
MEP	Multidisciplinary Expert Panel
MGRs	Marine genetic resources
MLS	Multilateral system
MOU	Memoranda of understanding
NBSAP	National Biodiversity Strategies and Action Plans
NBS LME	North Brazil Shelf LME
NGOs	Non-Governmental Organizations
NP	Nagoya Protocol on Access to Genetic Resources and the Fair and Equitable Sharing of Benefits Arising from their Utilization to the Convention on Biological Diversity
OECS	Organization of Eastern Caribbean States
OIE	World Organization for Animal Health
OSPAR	Convention for the Protection of the Marine Environment of the North-East Atlantic
PAME	Protection of the Arctic Marine Environment Working Group
PCA	Permanent Court of Arbitration
PIC	Prior informed consent
PMHs	Priority marine habitats
PPM	Processes and production method
Ramsar Convention	Convention on Wetlands of International Importance Especially as Waterfowl Habita
RSP	Regional Seas Programme

RTAs	Regional trade agreements
SAOs	Senior Arctic Officials
SAP BIO	Strategic action plan for biodiversity
SAs	Special areas
SBSTTA	Subsidiary Body on Scientific, Technical and Technological Advice
SDGs	Sustainable Development Goals
SDWG	Sustainable Development Working Group
SEUS LME	Southeast US Continental Shelf LME
SIDS	Small Island Developing States
SMTA	Standard Material Transfer Agreement
SPS Agreement	Agreement on the Application of Sanitary and Phytosanitary Measures
SNIJ	Sub-National Island Jurisdictions
STOs	Specific trade obligations
STRP	Scientific and Technical Review Panel
TBT Agreement	Agreement on Technical Barriers to Trade
TED	Turtle excluder device
TFEU	Treaty on the Functioning of the European Union
TK	Traditional knowledge
TPP	Trans-Pacific Partnership
TSD	Trade and sustainable development
TTIP	Transatlantic Trade and Investment Partnership
UN	United Nations
UNCLOS	United Nations Convention on the Law of the Sea
UNCSD	United Nations Conference on Sustainable Development
UNDP	United Nations Development Programme
UNEP	United Nations Environment Programme
UNEP CEP	Caribbean Environment Programme
UNEP-WCMC	United Nations Environment Programme World Conservation Monitoring Centre
UNESCO	United Nations Education, Scientific and Culture Organization
UNFCC	United Nations Framework Convention on Climate Change
UNGA	United Nations General Assembly
UNICEF	United Nations Children's Fund
UNOHRLLS	United Nations Office of the High Representative for the Least Developed Countries, Landlocked Developing Countries and Small Island Developing States
UPOV Convention	International Convention for the Protection of New Varieties of Plants
US NOAA	United States National Oceanic and Atmospheric Administration
VCLT	Vienna Convention on the Law of Treaties

WCMC	World Conservation Monitoring Centre
WCR	Wider Caribbean Region
WHC	World Heritage Convention
WHO	World Health Organization
WTO	World Trade Organization
WWF	World Wildlife Fund

Chapter 1
Introduction

Mar Campins Eritja and Teresa Fajardo del Castillo

Contents

1.1 Introduction 2
1.2 A General Framework of the Interactions Between the Convention on Biological Diversity and Other International Environmental Regimes 3
1.3 Presentation of the Book 6
References 11

Abstract In the development of the Convention on Biological Diversity (CBD) and the biodiversity-related conventions, and their adaptation to new challenges, the updating of scientific knowledge is playing an increasingly important role. This is particularly true of the decisions approved by the Conference of the Parties (CoP) to encourage the implementation of Strategic Plan for Biodiversity 2011–2020 and the Aichi Targets through the adoption of National Biodiversity Strategies and Action Plans.

Scientific knowledge has also played a major part in the assessment of the implementation and enforcement of these legal regimes at international and national levels. It has found a way to inspire law-making through soft and hard law instruments in order to adapt the CBD and the biodiversity-related conventions to new circumstances. Coordination between the CBD and those conventions and other multilateral agreements must be part of a strategy that is able to persuade States to accept change and reform in line with a scientific imperative rather than an exclusively economic one.

After presenting an outline of the current status quo, this introductory chapter describes the structure of the book, which is organized around eleven chapters that examine sectors or international regimes which interact with the CBD. This

---

M. Campins Eritja (✉)
Universitat de Barcelona, Faculty of Law, Barcelona, Spain
e-mail: mcampins@ub.edu

T. Fajardo del Castillo (✉)
Universidad de Granada, Faculty of Law, Granada, Spain
e-mail: fajardo@ugr.es

M. Campins Eritja, T. Fajardo del Castillo (eds.), *Biological Diversity and International Law*, https://doi.org/10.1007/978-3-030-72961-5_1

introduction also discusses the research questions that each chapter analyses and the methodology applied.

1.1 Introduction

Years before the adoption of the Convention on Biological Diversity in 1992 (CBD), many international treaties with a regional or sectorial perspective on biodiversity had already been adopted. Over time, many other instruments, both binding and non-binding, were approved to address new challenges to biodiversity posed by the loss of species, the fragmentation of ecosystems, the increasing exploitation of genetic resources or the prospect of exploiting resources in marine areas beyond national jurisdictions. Ever since its adoption, this umbrella convention, conceived as a way of ensuring the comprehensive protection of biodiversity and of fostering adaptation to new challenges, has served as a beacon for all the subsequent biodiversity-related conventions which have undergone many changes in their attempts to address new problems and to adapt to new scientific knowledge. With this vocation in mind, the Strategic Plan for Biodiversity 2011–2020, adopted by the CBD's Conference of the Parties (CoP) in 2011, proclaimed that it would "represent a useful flexible framework that is relevant to all biodiversity-related conventions" (Decision X/2). Now, when a new global biodiversity framework for 2030 is in progress, various alternatives are being explored to create

> synergies with other relevant multilateral environmental agreements and processes in particular through (...) the formalization and strengthening of mechanisms for collaboration between the three Rio conventions and the biodiversity-related conventions (CBD/WG2020/2/4, p. 54).

However, the ambition of a new global biodiversity framework for 2030 remains an open question. Opposing positions have emerged in the preparation of *Draft Zero*, which will be submitted to the next CoP in Kunming, China, in May 2021. Some State Parties argue that the post-2020 global biodiversity framework "should be limited to issues that are within the mandate of the Convention on Biological Diversity". Others argue that "the framework should be universal and should not undermine other conventions or objectives". Adopting an intermediate position, others still have suggested

> that there could be one set of headline targets with clear links to the CBD and another set of separate 'complementary' targets that could address biodiversity issues related to other processes (CBD/WG2020/2/4, p. 3).

In the meantime, practice shows that some of these biodiversity-related conventions have been brought into line with the Framework convention through the intervention of the United Nations Environment Program (UNEP), which has displayed an extraordinary capacity of communication between the different conventions and their institutional systems, most of which it supports as the secretariat. The recognition of the importance of the engagement of these other multilateral and regional environmental agreements in the preparation of the Post 2020 Global Biodiversity Framework led UNEP to consult multilateral environmental agreements (MEA) secretariats and governing bodies in order to promote cooperation and synergies (UNEA 2016). However, the commitment of these conventions to biodiversity governance is far from being harmonious; they have made it clear that their mandates and tasks differ not just in scope but also in normative intensity according to their foundational agreements. Significantly, their response to the current pandemic has also been different and in some cases divergent.

Despite these differences, the fight against the destruction of nature and the loss of biodiversity has become a common mission for them all. UNEP and the United Nations Environment Assembly (UNEA) have acted as catalysts for the joint efforts to support the Parties of the CBD and other MEAs so as "to strengthen their engagement in the negotiation process of the draft text (Zero draft) of the Post 2020 global biodiversity framework" (UNEA 2016) with the adjustments required by the pandemic.

1.2 A General Framework of the Interactions Between the Convention on Biological Diversity and Other International Environmental Regimes

The adoption of the CBD in 1992 and its objectives ushered in a new approach to biodiversity protection. The new circumstances of the planet in the Anthropocene era require a new vision that considers not only the gravity of the situation and the urgency of taking action, but also the contributions of scientific knowledge in setting the course of the changes needed (Tanaka 2019). Despite the current uncertainty, scientists and experts have spoken out loud and clear, but their proposals have made little impact on national plans and strategies, which continue to prioritize economic rather than scientific considerations. Notwithstanding the efforts to achieve the CBD's objectives through the implementation of the Aichi Targets, the adequate status of conservation of ecosystems and species cannot be achieved simply by designating new terrestrial and marine protected areas as foreseen in the Aichi Target 11.

Given that the stability of these MEAs and biodiversity-related conventions has been shaken by the current pandemic, it is as well to remember Handl's warning that environmental law "is no longer a single, well-defined product adopted with the expectation of stability for the foreseeable future. Rather, it is a fragile and temporary legal indicator in an institutionalised process where legal positions are under constant review and are susceptible to frequent and rapid alterations" (Handl 1990, p. 9). With its dependence on scientific knowledge and its evolution, international environmental law has in the case of the protection of biodiversity a clear example of the need for frequent and rapid alterations in order to adapt to scientific imperatives at both international and national levels. This duty to adapt to scientific knowledge also responds to a well-established obligation in international law: the obligation of due diligence on the part of States to fulfil their international commitments. Insofar as it requires States to keep abreast of technological progress and scientific developments, this concept is not static: it can be reflected in the development of specific principles such as prevention and, above all, precaution. However, as the current COVID-19 crisis has shown, international rules have not been able to translate the concern regarding the impact of human action on ecosystems and species into law or to reaffirm the principle of precaution as an indisputable precept. In fact, this principle is still considered merely as an approach in biodiversity-related conventions; for the States most reluctant to accept stronger commitments and who favour legal principles disguised in approaches and standards, it only serves as a limit not to err on the side of scientific knowledge.

With regard to scientific research and knowledge on biodiversity, first, it should be recalled that Principle 20 of the Stockholm Declaration in 1972 declared that

> Scientific research and development in the context of environmental problems, both national and multinational, must be promoted in all countries, especially the developing countries. In this connection, the free flow of up-to-date scientific information and transfer of experience must be supported and assisted, to facilitate the solution of environmental problems; environmental technologies should be made available to developing countries on terms which would encourage their wide dissemination without constituting an economic burden on the developing countries.

Certainly, this statement promoted the free flow of up-to-date scientific information to solve environmental problems, but it did not go so far as to assert the necessity of updating international law in a constant and dynamic process moving at the same speed as scientific knowledge. In fact, biodiversity-related conventions, like most MEAs, are updated in response to the state-of the art knowledge through a constellation of binding and non-binding instruments.

Thus, another feature common to all biodiversity-related conventions is that they are nurtured by scientific knowledge that establishes legal challenges and management principles to reach the "adequate status of conservation" taking into consideration the ecosystem approach or the interoperability principle that have developed the biodiversity legal regimes. In the wake of the global pandemic, it is of utmost importance to draw on scientific knowledge in search of an ecological imperative to save biodiversity.

The urgency to act also led States to turn to science at the Summit on Biodiversity of September 2020. They envisaged "accelerated biodiversity action, including through harnessing science, technology and innovation, strengthening capacity-building, and enhancing access and benefit-sharing, financing and partnerships for biodiversity" (Leaders' Pledge). Once again, as in Rio 1992, States have shown unwavering faith in the ability of science to provide solutions to environmental problems at minimal cost, when they state that "Science and technology, along with indigenous and local knowledge, can support transformations to sustainability". However, they also recognize that "the process and costs of critical biodiversity data collection, management and open access still need to be addressed" (Leaders' Pledge). Therefore, biodiversity-related conventions share the problem of undervaluing the information which scientific knowledge provides regarding their accomplishment and which should serve to enhance accountability regarding the enforcement and compliance of their rules, in particular those that are ambiguous or incorporate derogations and exceptions. The problem is especially serious in regard to the States Parties' demands for discretionary power to implement them, a power that is likely to undermine the possibility of achieving the conventions' goals and purposes.

Practice has shown, for example, that in the process of selecting areas of protection in accordance with the Aichi targets, the States Parties to the CBD have chosen them by adopting an anthropocentric approach that gives priority to the economic costs of action. Thus, as stated in the literature: "the general bias toward low-cost conservation and indifference to biodiversity represents a major limitation for global efforts to slow biodiversity decline under existing biodiversity conventions (Venter et al. 2018, pp. 131–132)".

Therefore, scientific reports have denounced that States define open categories of protection at the expense of biological diversity, prioritizing economic and cultural activities such as the increase of agricultural lands or licencing hunting and indiscriminate culling as biodiversity management tools. A new approach is proposed by the CoPs, which have concluded that scientific knowledge must help to avoid bias as well as abusive interpretation of their obligations or the use of derogations and exceptions to them that have eroded the achievement of their goals.

The Science-Policy Platform on Biodiversity and Ecosystem Services (IPBES) was established in 2012 as an independent intergovernmental body to act in the science-policy interface on biodiversity governance, in a way similar as the Intergovernmental Panel on Climate Change (IPCC) does for the climate change regime. Its 2018 Report raised awareness about biodiversity loss due to neglect and the overexploitation of natural resources by States as one of the features of the Anthropocene (IPBES 2019). However, since the beginning of the new millennium, the Global Environmental Outlook Reports and the Global Outlooks on Biodiversity, published annually by UNEP and the CBD, have sent the same message of urgency and distress—without eliciting a response from States with the capacity to change the course of the loss of biodiversity caused by overexploitation of resources, change of land use, pollution and climate change. These reports led academics and experts to conclude that "human governance systems at both the international and national levels have failed, and continue to fail, to prevent human activities undermining the integrity, vitality, health and functioning of the natural communities of the planet (Cullinan 2013, p. 95)".

The diagnosis then was that the failure to protect the environment was due to the lack of enforcement of environmental law. A decade later, the same diagnosis was issued in the UNEP Report on the Environmental Rule of Law, published in 2018. The postponed fifteenth CoP of the CBD was expected to offer the right answers to reverse these processes. The urgency of the situation led to the organization of the Summit on Biodiversity that took place in the General Assembly on 30 September

2020. Previous to this summit, a small group of global leaders adopted a Pledge for Nature to support the role of the United Nations in saving biodiversity in times of exacerbation of risks and damage due to the global pandemic. In both documents, of a non-binding nature but with important legal implications for the future, scientific knowledge was once again at the heart of the solutions.

In the future revision of the Aichi Targets and the adoption of a new Biodiversity Plan for 2030, the CBD Subsidiary Body on Implementation invites the Subsidiary Body on Scientific, Technical and Technological Advice to carry out a scientific and technical review of the goals and targets, updated in the light of its second meeting, as well as related indicators and baselines (CBD/WG2020/2/4 p. 1). This proposal reflects a recognition of the importance of scientific knowledge in the adoption of a new rule of law for biodiversity based on an ecological imperative for the implementation of suitable measures to address and reverse current trends.

One of the biggest problems facing the planet is the failure to attain the *Vision* foreseen in the Strategic Plan 2011–2020 and the Aichi Targets concerning value, conservation, restoration and wise use of biodiversity by 2050 (Decision X/2). The main challenge is how to achieve a balance between sovereignty and national interests and the goals of biodiversity conservation and protection. Given the gravity of the current biodiversity loss and the global pandemic, the decade ahead will challenge the objectives of biodiversity-related conventions, obliging them to confront both long-delayed problems and new ones that were predictable but were exacerbated by the pandemic.

The development of biodiversity-related conventions must be underpinned by scientific knowledge and this must become the imperative to regulate and to act in the face of the current challenges. Thus, biodiversity-related conventions have to go further to achieve “anything of significance for ecological sustainability” so as to be the groundnorm (Bosselmann 2013) for biodiversity protection or the biological imperative that supersedes the sovereignty interests when scientific knowledge clearly shows the ways and means. But scientific knowledge is also susceptible to manipulation; this is why the enhancement of international governance is decisive to preserve the much-needed expertise and capacity of assessment.

1.3 Presentation of the Book

This joint publication analyses certain international legal mechanisms through which the interaction between the CBD and other environment-related conventions takes place, in order to strengthen the complementary approaches that they can provide. This is done through an appraisal of selected cases that provide traceable instances of potential interactions between the CBD and other regimes that come about through a variety of measures characterized by different levels of regulatory intensity. The goals and principles of the CBD and the Strategic Plan for Biodiversity and the Aichi Targets have been mainstreamed in the constellation of biodiversity-related conventions, albeit with very different results and degrees of influence. The lack of strong links between these conventions is presented as one of the causes of the lack of effectiveness of biodiversity law. The fragmentation of the legal regimes and limited institutionalization are identified as the key factors.

From these premises, the specific aim of this book is to address the systemic challenges presented by this area, through a twofold approach: (a) the analysis of legal interactions between international biodiversity law and related international law that affect the conservation and sustainable use of biodiversity; and (b) the analysis of issues related to the governance of biodiversity from a functional, normative and geographical perspective in order to produce a crosscutting and holistic approach. In addition, the book uses regional case studies to strengthen a horizontal and joint approach of the various dimensions of the field of biodiversity, the different legal-institutional regimes present, and their implications.

In order to undertake this analysis, the book is divided into ten chapters (in addition to this introductory chapter) that examine different sectors of biodiversity regimes. In the first part, after introducing the general framework, the book presents the central principles that underpin biodiversity-related conventional law, the soft law that conveys targets for enforcement of the CBD, their structural, regulatory and implementation gaps and the systemic relations arising from national interests, and the role of scientific advisory bodies in biodiversity-related agreements. The second part addresses interactions that occur in specific conventional frameworks, such as the law of multilateral trade and global public health. Finally, the third part illustrates these issues by analysing four case studies focusing on the challenges for sustainability and marine biodiversity in the case of small islands, the Arctic Ocean, the Caribbean and the Mediterranean Sea.

The various chapters offer a crosscutting analysis that focuses especially on the legal development of biodiversity-related conventions and their adaptation to the new circumstances and advances in scientific knowledge. The CBD and its institutions have been the catalyst for cooperation between legal regimes and a source for common interpretation, and for providing technical and scientific information relating to the new trends such as the protection of marine areas beyond national jurisdictions. Moreover, the CBD and its principles have acted as drivers of legal development of the different biodiversity-related conventions such as the Barcelona Convention for the Protection of the Marine Environment and the Coastal Region of the Mediterranean and the Agreement on the Conservation of Cetaceans of the Black Sea, Mediterranean Sea and Contiguous Atlantic Area. Despite the fact that most of objectives of the Strategic Plan for Biodiversity and its Aichi Targets have not yet been achieved, their commitments must be irrevocably renewed in order to deal with COVID-19 and the grave problems facing the planet and humanity.

Because all biodiversity-related conventions suffer from the common problem of lack of implementation and compliance, most efforts to develop innovative legislation have sought to find new tools and mechanisms to overcome the general ineffectiveness of environmental law. A wide range of binding and non-binding instruments have been designed to persuade States to accept the commitments necessary to achieve the objectives being targeted, but they lack the crucial support of reporting mechanisms for promoting and monitoring compliance, so that, in the event of violations, sanctions can be imposed at international level. Principles and approaches, guidelines and regional regimes have sought to adapt the objectives of the CBD to a planet under threat from old and new hazards such as COVID-19. However, compliance depends on national policies and regimes, as in the case of the designation of terrestrial and marine special protected areas, and unfortunately their non-binding nature, and weak governance that does not go beyond the implementation of reporting and monitoring mechanisms, leave non-compliance unpunished. CITES, one of the exceptions, is upgrading its role in the current fight against possible vectors of contagion.

The capacity to adapt to new legal trends and problems will certainly mark out the path for future research. Adaptation will be the key approach to the development of legal instruments of all kinds and of international and national policies. Thus, UNEP has called for a holistic adaptation strategy that will create synergies between different regimes: in particular, incorporating the use of ecosystem services to guarantee the presence of healthy ecosystems in order to reduce the impact of climate change, and preserve biodiversity (UNEP 2019).

The first part of the book comprises three chapters.

After this brief introduction, the second chapter (*Principles and approaches in the Framework Convention on Biological Diversity and other biodiversity-related conventions in the post-2020s Scenario)*, by Teresa Fajardo del Castillo, highlights the momentum created by the imminent revision of the Strategic Plan for Biodiversity 2011–2020 and the Aichi Targets, in order to reconsider the role that the principles and approaches can play in solving the perennial problems of the lack of enforcement of, and compliance with, international environmental law. It explores the need to "update the basic principles of the framework convention in order to find a response to changing circumstances and expanding scientific knowledge".

José Juste Ruiz presents the third chapter (*Gaps in international biodiversity law and possible ways forward*), which focuses on "the existing gaps in international biodiversity law and explores possible ways forward to a more efficient and integrated legal regime". Juste Ruiz's analysis indicates the existence of structural, regulatory and implementation gaps that seriously undermine the efficiency of the CBD regime. He argues that "the preservation of the legal autonomy of each convention regime" is in line with the promotion of the particular interests of States to the detriment of the institutionalization of biodiversity problem-solving. Thus, "the fragmentation of the legal regime generates functional overlapping and a duplication of effort and it may be a source of legal gaps". Addressing the gaps in the implementation and enforcement and the harmonization of reporting requirements are currently on the agenda of several biodiversity conventions.

Roy Gardner sheds light on a subject as complex as the work of the scientific advisory bodies of the most important biodiversity-related conventions in Chap. 4 (*Opportunities and challenges for synergies across biodiversity-related conventions in the context of human health and zoonotic diseases: the role of scientific advisory bodies*). It examines how the workload of these agencies has steadily grown, pushing their budgets to the limit. It also shows how these bodies have coordinated in recent years and how this cooperation can be enhanced in the times of the COVID-19, in particular, inside the IPBES.

Part two comprises another three chapters.

The fifth chapter (*Conservation and sustainable use of biodiversity in the international regulation of trade in goods*), by Xavier Fernández-Pons, is devoted to examining how the objectives of biodiversity conservation and sustainable use have been addressed in the multilateral trading system and other trade agreements such as those concluded by the EU with third countries. It analyses the emblematic cases of trade-related biodiversity measures raised through the WTO's dispute settlement system. But, especially, it examines new proposals in the conventions to address the problem of the most harmful subsidies in the fishing sector, which are explicitly linked to Target 14.6 of the Sustainable Development Goals, concluding that the systemic interconnections being established between the negotiations at the WTO and the SDGs serve "to overcome the traditional fragmentation between various sectors of the international legal system".

In the sixth chapter (*Is everything marketable in international trade? Public health issues in the international trade of wildlife*), Josep Maria de Dios Marcer analyses the legal movement of species as part of international trade, focusing on the aspects that require special attention in the times of the pandemic. The chapter studies old and new problems related to the CITES Convention, as well as the possible "impact on public health of the importation of endangered species, given that they may carry pathogens and introduce them into their new environment". At present, from the regulatory point of view, the responses articulated at international and European levels are dispersed and need greater cooperation between the different treaty regimes and better implementation.

In the seventh chapter (*Access to and benefit-sharing of genetic resources: Rationale, review and prospects*), Xavier Pons Rafols analyses the scope and limits of the access and benefit sharing (ABS) system in the international biodiversity regime and the specific multilateral ABS regime on plant genetic resources for food and agriculture. In a prospective review, the analysis looks at the main challenges facing the internationally established ABS mechanism, in particular, the digitization of genetic sequence information that might be a "crucial and truly disruptive element" since it will make it not necessary to have access to genetic resources. It also explores the on-going work on an ABS mechanism for the future agreement on marine biodiversity in areas beyond national jurisdictions and the possibility of establishing an ABS mechanism for marine genetic resources.

Part three comprises four chapters.

Francesco Sindico focuses his analysis on the role of sub-national island jurisdictions (SNIJs) as meaningful biodiversity actors in the eighth chapter (*Sub-National Island Jurisdictions and Biodiversity Protection: Moving towards Greater Resilience and Sustainability*). This chapter discusses how SNIJs may rely on the CBD to become more resilient and sustainable. It focuses on the Åland Islands in Finland as an example of how the SDGs and the CBD might contribute to the shaping and development of the official biodiversity policies of a SNIJ.

The ninth chapter (*The Arctic Ocean: Ecosystem approach in a context of extreme vulnerability*), by Mar Campins Eritja, focuses on the influence of the CBD in the protection of the Arctic Ocean's biodiversity. It examines the role played by the CBD's ecosystem approach to address the problems of management of the vulnerable Arctic Ocean's marine ecosystems, in particular, informing the policies and legal measures adopted by the institutions of the Artic governance. However, as Campins Eritja notes, "the management of the Arctic Ocean based on an ecosystem approach demands a holistic vision that international law so far seems ill-equipped to provide". Thus, she proposes

> a special legal framework including the prioritization of objectives and the effective use of the scientific knowledge available in order to develop and make the ecosystem approach operative.

In the tenth chapter (*Governance of the Caribbean Marine Biodiversity: Complex Challenges for a Complex Region*), Luis Enrique Rodríguez-Rivera maps the multilevel governance of biodiversity in the Caribbean "which is based on a framework made up of an intricate, overlapping web" of international and regional agreements and institutions. The effectiveness of the Caribbean marine biodiversity governance framework is undermined by different problems that are critically examined in this chapter. Their solution depends on global help and investments

> to repair the environmental, political and democratic degradation in the region.

Guillaume Futhazar, in the eleventh and last chapter (*The Convention on Biological Diversity and the Mediterranean Sea: A beacon for international environmental governance in the region*) examines the overall influence of the CBD on the development of the legal regimes and governance of the Mediterranean Sea through normative diffusion and informal legal ways. However, "the Mediterranean is but another unfortunate example of the general ineffectiveness of environmental law". Nonetheless, Futhazar criticizes the fact that its influence is still weak and not capable of real change, since the interactions with other legal regimes such as the United Nations Convention on the Law of the Sea through the designation of Mediterranean Ecological or Biologically Significant Marine Areas have no direct legal consequences.

The book is one of the final outcomes of the research project "Biological Diversity and International Law: Emerging Issues and Trends, Interaction of Legal Regimes and Key Policy Challenges", (*Diversidad Biológica y Derecho Internacional: Nuevos Problemas y Tendencias, Interacción de Regímenes Jurídicos y Principales Retos Normativos*" (DER2017-85406-P), funded by the Spanish Ministerio de Ciencia, Innovación y Universidades (now *Ministerio de Ciencia e Innovación*), which have been pursued between 2017 and 2020 by a group of researchers and professors in Public International Law who share a common interest: the legal protection of the environment. Therefore, from a methodological viewpoint, the methods of legal science have dominated the preparation of each chapter, reflecting the legal specialization of each author. For this reason, each analysis brings together the various relevant elements of international law, adopting a dual perspective that addresses not only the rules and institutions but also the processes for their formulation and execution within the framework of each area.

We should note that the drafting of the chapters of this joint publication was completed in December 2020. The areas analysed in the work are particularly dynamic, and we are well aware that there may be a certain time lag between the completion of our study and its final publication. However, we believe that as a collective publication, it offers an approach that allows us to address the major legal issues, beyond specific time constraints and the rapid development of current events.

Finally, we also note that the opinions expressed in this work correspond exclusively to its authors. We thank all the anonymous colleagues and experts who have contributed to the publication of this book through their valuable inputs to the scientific peer review of the originals. Their comments, always pertinent and constructive, have greatly enriched our work. We would also like to extend our thanks to the editors of Springer Nature for their trust and support throughout this process. We hope very much that we will prove worthy of it.

References

Bosselmann, K. (2013). Grounding the rule of law. In C. Voigt (Ed.), *Rule of law for nature* (pp. 75–93). Cambridge: Cambridge University Press.

CBD. (2020). Report of the Open-Ended Working Group on the Post-2020 Global Biodiversity Framework on its Second Meeting, CBD/WG2020/2/4.

Cullinan, C. (2013). The rule of nature's law. In C. Voigt (Ed.), *Rule of law for nature* (pp. 94–108). Cambridge: Cambridge University Press.

Decision X/2 Adopted by the Conference of the Parties to the Convention on Biological Diversity at its Tenth Meeting: The Strategic Plan for Biodiversity 2011-2020 and the Aichi Biodiversity Targets, UNEP/CBD/COP/DEC/X/2, 29 October 2010.

Handl, G. (1990). Environmental security and global change: The challenge to international law. *Yearbook of International Environmental Law, 1*(1), 3–33.

IPBES. (2019). Summary for policymakers of the global assessment report on biodiversity and ecosystem services of the intergovernmental science-policy platform on biodiversity and ecosystem services. In S. Díaz, J. Settele, E. S. Brondízio, H. T. Ngo, M. Guèze, J. Agard, A. Arneth, P. Balvanera, K. A. Brauman, S. H. M. Butchart, K. M. A. Chan, L. A. Garibaldi, K. Ichii, J. Liu, S. M. Subramanian, G. F. Midgley, P. Miloslavich, Z. Molnár, D. Obura, A. Pfaff, S. Polasky, A. Purvis, J. Razzaque, B. Reyers, R. R. Chowdhury, Y. J. Shin, I. J. Visseren-Hamakers, K. J. Willis, & C. N. Zayas (Eds.), *IPBES secretariat* (p. 56). Bonn, Germany. https://doi.org/10.5281/zenodo.3553579

Tanaka, T. (2019). Governance for protected areas "beyond the boundary"—A conceptual framework for biodiversity conservation in the Anthropocene. In M. Lim (Ed.), *Charting environmental law futures in the Anthropocene* (pp. 71–79). Singapore: Springer.

UNEA. (2016). *Promoting the engagement of MEAs in the negotiation process of the zero draft of the Post 2020 global biodiversity framework.* https://www.unenvironment.org/news-and-stories/story/promoting-engagement-meas-negotiation-process-zero-draft-post-2020-global

Venter, O., Magrach, A., Outram, N., Klein, C. J., Possingham, H. P., Di Marco, M., et al. (2018). Bias in protected-area location and its effects on long-term aspirations of biodiversity conventions. *Conservation Biology, 32*, 127–134.

Part I
Principles Inspiring Biodiversity-Related Conventional Law and Governance

Chapter 2
Principles and Approaches in the Convention on Biological Diversity and Other Biodiversity-Related Conventions in the Post-2020 Scenario

Teresa Fajardo del Castillo

Contents

2.1 Introduction 16
2.2 The Legal Nature and Functions of Principles and Approaches on Biodiversity 18
2.3 The General Principles of International Law and the Principles of International Environmental Law in Biodiversity-Related Conventions 22
2.4 Specific Principles of the Framework Convention on Biological Diversity and the Binding and Non-Binding Instruments That Develop It 28
2.5 Principles for the Post-2020 Scenario According to the Summit on Biodiversity and the Zero Drafts of the CBD COP 15 31
2.6 Conclusion 32
References 33

Abstract This chapter will examine principles and approaches in the Convention on Biological Diversity (CBD), other biodiversity-related conventions and soft law from a post-2020 perspective. The global COVID-19 pandemic and the imminent revision of the Strategic Plan for Biodiversity 2011–2020 and the Aichi Targets have created the momentum to reconsider the role that these principles and approaches can play in solving the perennial problems of lack of enforcement of, and lack of compliance with, international environmental law. Principles and approaches are of particular importance in addressing these problems, as they inform both the design and the implementation of National Biodiversity Strategies and Action Plans necessary to achieve the objectives of the CBD and other biodiversity-related conventions, taking into account the capacities of States and their national circumstances.

These principles and approaches have systemic, substantive and procedural functions in the fragmented legal mosaic of biodiversity, inspiring its future legal development, interpretation and norm-application at both international and domestic levels. Existing and emerging principles and approaches in biodiversity-related

T. Fajardo del Castillo (✉)
Universidad de Granada, Faculty of Law, Granada, Spain
e-mail: fajardo@ugr.es

M. Campins Eritja, T. Fajardo del Castillo (eds.), *Biological Diversity and International Law*, https://doi.org/10.1007/978-3-030-72961-5_2

conventions are also developed through soft law instruments that adapt to new challenges such as the COVID-19 pandemic. The state-of-the-art principles for the conservation of biological diversity in the times of the coronavirus provide legal tools for facing an ecological crisis of an unprecedented scale. Biodiversity governance should encourage interoperability between biodiversity-related regimes, which run the risk losing synergies at a time when they are needed more than ever before.

2.1 Introduction

The principles and approaches of international environmental law play multiple roles in biodiversity-related conventions, and vary widely in terms of their legal intensity and ambition. Now, they may serve to adopt a new approach to the exercise of sovereignty in order to attain the goal of conserving biological diversity as a common concern of humankind, as was stated in the preamble of the Convention on Biological Diversity (CBD). The global COVID-19 pandemic and the imminent revision of the Strategic Plan for Biodiversity 2011–2020 and the Aichi Targets have created the momentum to reconsider the role that these principles and approaches can play in solving the perennial problems of lack of enforcement of, and lack of compliance with, international environmental law. However, the rules and measures necessary to accomplish the CBD goals as formulated in its Article 1: "the conservation of biological diversity, the sustainable use of its components, and the fair and equitable sharing of the benefits arising out of the utilization of genetic resources", have to be adapted and implemented at national level. The principles and approaches associated with these objectives, such as the conservation principle, sustainable use, the principle of environmental impact assessment and the ecosystem approach, have made it easier for scientific imperatives to permeate the development and adaptation of biodiversity regimes to new circumstances at a faster rate than conventional standards, which are held back by the difficulty of achieving consensus in treaty making.

This chapter will examine the principles and approaches incorporated in the CBD and in other biodiversity-related conventions and in soft law from a post-2020 perspective. The governing bodies of the CBD and other biodiversity-related conventions are currently referring to these principles and approaches in relation to the revision of the Strategic Plan for Biodiversity 2011–2020 and the Aichi Targets. Furthermore, the current situation caused by the pandemic will test the ability of principles and approaches in the new post-2020 period to adapt to the new challenges and to contribute to solutions in decision- and policy-making that take into account the capacities of states and their circumstances. Until now, however, to quote Krämer:

> the loss of biological diversity is an accepted global phenomenon which is not seriously disputed. Yet when it comes to conflicts between economic interests – logging, mining, the building of infrastructure, fishing, agricultural activities or urbanization – the natural environment is almost always the loser. No environmental principle – sustainable development or sustainable use of natural resources, environmental impact assessments or the prevention principle – has succeeded in causing a significant reduction or a stop to the loss of biological diversity. (Krämer 2018, p. 7).

The Strategic Plan for Biodiversity 2011–2020 and the Aichi Targets were global policies formulated to achieve the goals of the CBD through the adoption of national rules, plans and strategies. They are currently being evaluated and reviewed in order to establish why their objectives have not been achieved. Principles and approaches will be used to evaluate both the design and the implementation of the National Biodiversity Strategies and Action Plans (NBSAPs) required to pursue the objectives of the CBD and other biodiversity-related conventions. Both the processes and the results they were intended to achieve are being evaluated in the light of scientific developments and new indicators that measure compliance. In this scenario, the principles and approaches of the CBD and biodiversity-related conventions have also been adapted in line with new scientific developments and will be translated into new formulations showing a progressive development of international biodiversity law beyond the margins of the treaty-making processes, using soft law instruments to better guide and clarify the weak and ambiguous rules of the CBD which were needed to achieve almost universal participation by States—with the exception of the US and its decision to depart from the biodiversity regimes and the principles they uphold.

The principles and approaches of the biodiversity-related conventions have systemic, substantive and procedural functions in the fragmented legal mosaic of biodiversity, underpinning its impending legal development, interpretation, and norm-application at both international and domestic levels. The existing and emerging principles and approaches of the biodiversity-related conventions are also developed through soft law to enable adaptation to new challenges such as the COVID-19 pandemic, providing a rapid response that will require additional support from States. The state-of-the-art principles for the conservation of biological diversity in the times of the coronavirus will provide the legal tools for facing an ecological crisis of an unprecedented scale.

The following sections will focus first on the legal nature of the principles and approaches of biodiversity-related conventions and those formulated in the binding and non-binding instruments that developed these conventions, subsequently by their institutional bodies of representation. The next section will then examine the general principles of international law and international environmental law underlying the CBD and the other biodiversity-related conventions from the point of view of their ability to adapt to their specific goals and characteristics. The fourth section will examine the principles specific to the biodiversity-related conventions, and in particular those adopted by the Conference of the Parties (CoP) and the governing and technical bodies, with a view to updating the basic principles of the framework convention in order to find a response to changing circumstances and expanding scientific knowledge. A final section on the state-of-the-art principles for the conservation of biological diversity in the times of coronavirus will address some of the responses offered by multilateral environmental agreements in partnership with the CBD. The final remarks will make recommendations from the perspective of the review of the Strategic Plan for Biodiversity 2011–2020 and the Aichi Targets for the next decade.

2.2 The Legal Nature and Functions of Principles and Approaches on Biodiversity

An analysis of principles and approaches requires an examination of their legal nature and functions so that they can be distinguished from the conventions from which they are drawn in the legal system governing biodiversity. Whether or not they are considered as a source of law or as the underlying rationale behind treaties and customs, -a debate which it is not up to us to settle here (Kohen and Schramm 2017), in the sectorial regime of biodiversity, the principles and approaches adopted in conventional instruments and soft law are undoubtedly the legal foundations behind norms and policies. Their legal value when incorporated into binding and non-binding instruments and into global and national policies also requires further analysis depending on the function they are called upon to perform. As Krämer says, principles "in this sense [. . .] represent the ideal bridge between ideals and duties, forging the link between the 'morality of aspiration' and the 'morality of duty', although they may be legally binding" (Krämer 2018, pp. 3–4). Furthermore, to quote Koester,

> leaving aside that a number of the principles are customary international law, the principles should be regarded as complementary to or supplementing biodiversity treaties (Koester 2018, p. 551).

The reason why principles and approaches have become complementary to the biodiversity treaties is that problems in reaching consensus on sovereignty-sensitive areas have led to the replacement of conventional standards by soft law and by principles and approaches enshrined in the decisions and resolutions which have been adopted by the governing bodies and the Conference of the Parties and now underpin both global and local strategies.

In the case of the CBD, its nature as a framework agreement determines a dual role for principles and approaches, because this type of treaty, as defined by Kiss ". . . sets out the principles that should serve as a basis for cooperation between States Parties in a given field, while leaving it to them to define, in separate agreements, the modalities and details of the cooperation, providing, where suitable, for one or more appropriate institutions for that purpose" (Kiss 1993, p. 793). Therefore, these principles and approaches serve to structure a legal regime in the making (Boisson de Chazournes and Maljean du Bois 2016, p. 3) with a forward-looking vision that is open to change and to new scientific knowledge (Juste 1997). The general nature of the rules and the weak legal content of framework conventions such as the CBD also mean that principles and approaches play a major role in the process of transposition and implementation at the national level. Principles and approaches are thus the fabric of the current and future rule of law for biodiversity since they inform the definition, the interpretation and the development of its rules at international and domestic levels. The United Nations Environment Programme (UNEP) has a major role to play in weaving the fabric of this rule of law, as it has made widespread use of a common set of principles and approaches in the field of biodiversity which has been developed further by the governing bodies of biodiversity-related conventions.

The difference between principles and approaches lies in the existence of consensus in their definition, content and scope. Approaches are principles *in statu nascendi*. Most of the principles which have been rejected during the negotiations by certain States and have been the subject of formal objections become approaches in the text of the treaty or in the subsequent soft law that complements it by decision of the States Parties, as was the case during the preparatory negotiations for the "draft text of an agreement under the United Nations Convention on the Law of the Sea on the conservation and sustainable use of marine biological diversity of areas beyond national jurisdiction" (BBNJ). In the latest draft of this convention, possible alternatives regarding principles and approaches still appear in brackets, pending the States Parties' future pronouncement on the ambition of their commitments. At the start of the negotiations, the possibility was discussed of recovering the principles and approaches that were not incorporated in the resulting treaty in a soft law instrument, with the potential for regulation in the future, and despite the sensitive nature of some of them.[1] Thus, Article 5 of the draft of the BBNJ agreement will deal with "General [principles] [and] [approaches]" and says:

> In order to achieve the objective of this Agreement, States Parties shall be guided by the following:
>
> [(a) The principle of non-regression;]
>
> (b) [The polluter pays principle] [The endeavour to promote the internalization of environmental costs and the use of economic instruments, taking into account the approach that the polluter should [, in principle,] bear the cost of pollution, with due regard to the public interest and without distorting international trade and investment];
>
> [(c) The principle of the common heritage of mankind;]
>
> [(d) The principle of equity;]
>
> (e) The precautionary [principle] [approach];
>
> (f) An ecosystem approach;
>
> [(g) An integrated approach;]
>
> (h) An approach that builds ecosystem resilience to the adverse effects of climate change and ocean acidification and restores ecosystem integrity;

[1]For example, the principle of adjacency no longer appears among them.

(i) The use of the best available [science] [scientific information and relevant traditional knowledge of indigenous peoples and local communities];

(j) The non-transfer, directly or indirectly, of damage or hazards from one area to another and the non-transformation of one type of pollution into another". (GA 2019, p. 7).

Their final definition as an approach or a principle will limit their legal scope, but not their usefulness as standard-setters. Thus principles and approaches are here to stay, as witnessed by the persistent recognition of the role they play in treaty and policy-making as well as in the case law of international and domestic courts. However, their soft nature may lead to the application of general principles of international law and international environmental law that might take precedence over more specific principles which are ambiguously formulated, as in the case of the CBD principle on environmental impact assessment in its Article 14,[2] whose application has been extended to different biodiversity-related conventions; however, as shown in the case Nicaragua v. Costa Rica, the International Court of Justice applied the general principle of environmental impact assessment instead of the principle as formulated in the CBD because it merely requests of contracting parties that they apply it "as far as possible and as appropriate" (Sands et al. 2018, p. 393).

Moreover, specific principles for the protection of biodiversity are often embedded in annexes and protocols, and in decisions and resolutions adopted by the CoPs and governing bodies that have developed the legal framework for the protection of biodiversity further, with both substantive and procedural functions that apply not just to the soft and treaty-making processes but also to national implementation and compliance processes as guidance. Thus, the CBD CoP, with the support of the Ad Hoc Open-ended Working Group on Review of Implementation of the Convention, has played a constituent role when adopting further rules and policies inspired by principles – for example, when the CBD CoP 10 adopted Decision X/2 approving the Strategic Plan for Biodiversity 2011–2020 and the Aichi Targets which established that, in order to achieve their goals, the COP and the States Parties will apply "the principle of adaptive management through active learning" to allow "new guidance [to be] informed by the experience of Parties in implementing the Convention" (Decision X/2). The importance of principles and approaches in assessing the results of the implementation and enforcement of biodiversity legislation and action plans has also been highlighted, demonstrating how States can undermine the objectives pursued by the binding and non-binding instruments referring to biodiversity due to their national interests which prioritize economic concerns over environmental ones. Thus, the application of the ecosystem approach has shown that the Aichi Targets have not been achieved, either quantitatively or qualitatively.

---

[2]In its first paragraph, letter a, Art. 14 of the CBD states:

1. Each Contracting Party, as far as possible and as appropriate, shall:

(a) Introduce appropriate procedures requiring environmental impact assessment of its proposed projects that are likely to have significant adverse effects on biological diversity with a view to avoiding or minimizing such effects and, where appropriate, allow for public participation in such procedures; (. . .).

Not only has the designation of terrestrial and marine protected areas failed to meet the goal of increasing the share of the globe under protection, but those that have been designated often lack economic or biological value, or are located in such remote areas that adequate conservation measures are not guaranteed, even if they were adopted in the framework of binding instruments such as the EU Birds and Habitats Directives. Moreover, most areas do not meet the criteria for adequate status of conservation due to poor compliance with regimes and the abusive use of authorizations and derogations. These problems must be resolved if principles and approaches are to be used in the "identification of areas [requiring protection]" as stipulated in the first paragraph of Article 16 of the draft of the BBNJ agreement:

> 1. Areas requiring protection through the establishment of area-based management tools, including marine protected areas, shall be identified on the basis of the best available [science] [scientific information and relevant traditional knowledge of indigenous peoples and local communities], the precautionary [approach] [principle] and an ecosystem approach.

In other matters, biodiversity-related principles and approaches resolve systemic tensions in the complex and fragmented international legal system for biodiversity at the heart of which is the CBD by building links to other sectorial regimes with which they intersect (Fajardo 2018). This is clearly the case between the CBD and the United Nations Convention on the Law of the Sea (UNCLOS) and regional regimes of marine protection as well as the legal instruments in the field of human rights such as those protecting indigenous peoples and local communities (Morguera 2018). Now, during the pandemic, biodiversity-related conventions are also clearly linked to human health, as acknowledged by the principles adopted by UNESCO in its "One Health" initiative.

Therefore, in the process of the preparation of a new strategy for 2030, in the Zero Draft written by the Open-ended Working Group on the Post-2020 Global Biodiversity Framework the Parties of the CBD stated that "the framework should provide concrete guidance, commitments and mechanisms to ensure that synergies with the other Rio conventions and the 2030 Agenda for Sustainable Development are dramatically strengthened" (CBD/WG2020/2/4, p.52). Synergies will be sought with multilateral environmental agreements (MEAs) and not only with the Rio conventions, so the idea of mainstreaming biodiversity into all relevant MEAs informs the Zero Draft, from conventions on forests to conventions on chemical and waste processes. (CBD 2020, p. 23).

2.3 The General Principles of International Law and the Principles of International Environmental Law in Biodiversity-Related Conventions

Article 3 of the CBD is a canonical reference to the concept of sovereignty as consecrated in the UN Charter and to the principles underlying international environmental law as well. It says:

> States have, in accordance with the Charter of the United Nations and the principles of international law, the sovereign right to exploit their own resources pursuant to their own environmental policies, and the responsibility to ensure that activities within their jurisdiction or control do not cause damage to the environment of other States or of areas beyond the limits of national jurisdiction.

Thus, in a treaty provision under the title of 'principle', this Article 3 codifies the formulas adopted in Principle 21 of the Stockholm Declaration on the Human Environment and Principle 2 of the Rio Declaration on Environment and Development, which over time, and with the adherence of States, became customary law. However, the formula used is the original version of Principle 21, not the one finally introduced in Rio, where, at the request of developing countries, the "developmental" scope of the policies was added. This does not mean that the relationship between sovereignty and the protection of biological diversity has been clarified further, despite the fact that Article 4 defines the jurisdictional scope of the powers and responsibilities of States according to their correspondence with the material scope of the CBD[3] or the fact that Article 15 regarding access to genetic resources again makes reference to "the sovereign rights of States over their natural resources" or that "the authority to determine access to genetic resources rests with the national governments and is subject to national legislation". These references to the Charter and the principles of international law are a reminder that at the root of any environmental regime is the sovereignty of States over their natural resources (Willmore 2017), as well as the responsibility not to harm, i.e., the functional imperative to prevent environmental damage. Thus, the principle of sovereignty envisaged from the perspective of sustainability and biodiversity protection requires the conversion of State power over biodiversity into State responsibility. However, States do not act as guardians or trustees of biodiversity, despite the calls in climate litigation, whose principles must be applied to defend biodiversity before domestic courts and government offices. Nor have the rules on international responsibility for

[3]Article 4 on Jurisdictional Scope says:

> Subject to the rights of other States, and except as otherwise expressly provided in this Convention, the provisions of this Convention apply, in relation to each Contracting Party:
>
> (a) In the case of components of biological diversity, in areas within the limits of its national jurisdiction; and
> (b) In the case of processes and activities, regardless of where their effects occur, carried out under its jurisdiction or control, within the area of its national jurisdiction or beyond the limits of national jurisdiction.

damage to biological diversity shown any evolution, even though some development would be expected from a regulatory regime defined as being in the common interest of humankind and thus possessing the status of *erga omnes*. However, in the recent cases on Certain Activities Carried Out by Nicaragua in the Border Area (Costa Rica v. Nicaragua) Compensation Owed by the Republic of Nicaragua to the Republic of Costa Rica, the International Court of Justice gave the possibility to other States Parties to biodiversity-related conventions to intervene in the process as witnesses and experts under its Rule 43.[4]

Also a principle of international law, the principle of international cooperation applied to environmental protection includes the duty to promote the negotiation of treaties and other international instruments for this purpose. In response to calls from other States or international bodies, States must cooperate in good faith in the negotiation of such instruments for the protection of the environment, although it must be stressed that this obligation does not imply a parallel duty to conclude; thus, the negotiations are not legally bound to result in a treaty, when the interests involved have not been sufficiently accommodated (Juste 1997, p. 71). Thus, in the case of the protection of biological diversity, international cooperation has for the most part been legally articulated through political and non-binding instruments, which have established a horizon of legal principles and expectations rather than a fledgling legal regime. Furthermore, this cooperation has been characterized by a particularism that has led to fragmentation due to the specific sectors that are regulated, as well as the different legal systems they form, in spite of the existence of common principles (Van Asselt 2012).

The principle of common but differentiated responsibilities in the field of biological diversity has transformed the principle of reciprocity—the basis of general international law and other treaties—according to national circumstances and the capacities of States. Adaptation to national circumstances and capacities is required by the different biodiversity-related conventions, such as the Convention on International Trade in Endangered Species of Wild Fauna and Flora (CITES), the Convention on Migratory Species, the Bern Convention on the Conservation of European Wildlife and Natural Habitats and the Ramsar Convention on Wetlands of International Importance. The legal asymmetry reconciling the approaches of countries of the south and the north is formulated in Principle 7 of the Rio Declaration on Environment and Development, and further developed in its Principle 11 to modulate its interpretation to take due account of two points: first, that when States enact effective environmental legislation, "environmental standards, management objectives and priorities should reflect the environmental and developmental context to which they apply", and second, that "standards applied by some countries may be inappropriate and of unwarranted social and economic cost to other countries, in

[4]Thus "[p]ursuant to the instructions of the Court under Article 43 of its Rules, the Registrar addressed the notifications provided for in Article 63, paragraph 1 to call to witnesses and experts, of the Statute, to States Parties to the Ramsar Convention, to the 1992 Convention on Biological Diversity and to the 1992 Convention for the Conservation of the Biodiversity and Protection of Priority Wilderness Areas in Central America" (ICJ 2015, para. 12).

particular developing countries" (Principle 11). In the case of biodiversity, both demands are met through a bottom-up approach to implementation that differs according to the context and capacities of States (Leal-Arcas 2011; Campins Eritja 2015), as reflected in the nationally determined commitments and contributions that have been developed further in the biodiversity-related conventions through the National Biodiversity Strategies and Action Plans. This configuration of the biodiversity regime has raised the question of whether international rules of conservation and on sustainable use are constraining national sovereignty or providing a protected space for national action (Willmore 2017, p. 31).

In the consultations prior to the CBD CoP 15, the formula of voluntary commitments has been further addressed in order to explore possible ways to improve implementation and enforcement. Thus, the consultation on "options for strengthening implementation of the post-2020 global biodiversity framework, including means of fostering commitments and building political momentum, as well as on the need for, and modalities of, voluntary commitments" (Notification 2018-063) has received a positive response from many States willing to launch voluntary commitments to meet biodiversity challenges from Parties (individually or in coalition) and stakeholders, including the private sector, to strengthen implementation and achievement of targets (EU 2019).

Regarding the principles of international environmental law, as Koester noted,

> it is probably a common trend of biodiversity treaties that they relatively rarely contain explicit provisions on 'classical' principles like the principles of prevention, precaution, environmental impact assessment, integration, and polluter-pays (...). The reason is probably that biodiversity treaties are normally not regulating activities, sources of pollution, and other forms of interferences threatening to wildlife, habitats or ecosystems. In that sense the treaties rely on other international treaties addressing such threats, and which may be considered as the main sources of environmental principles (Koester 2018, p. 540).

Furthermore, it is another characteristic trend of the CBD that soft law instruments have been the tools used by governing bodies to adapt those classic principles and approaches and to incorporate the new ones in decisions of such political importance as the Strategic Plan for Biodiversity 2011–2020 and the Aichi Targets, which proposed the initiatives and actions to be carried out at the domestic level in accordance with both substantive and procedural principles. Thus the principle of public participation in decision-making and access to justice are addressed not just in the Aarhus Convention or in the new Escazú Convention, but in the Aichi Targets and the new revisited targets for the post-2020 Global Framework for Biodiversity as well. This principle, which appeared in Article 14 of the CBD (Impact Assessment and Minimizing Adverse Impacts), has been submitted to States' appraisal to decide to whether it was "appropriate" to "allow for public participation in such procedures". However, the process of ratification of the Escazú Convention, from which some States such as Chile have dissociated themselves and others such as Brazil have frozen their commitment, is symptomatic of the importance of access to information, public participation in decision-making and access to justice in environmental matters, and also of the States' rejection of any effective control over the rule of law in environmental matters, particularly in the field of biodiversity. That is why the specific proposal put forward in the Zero Draft to be presented to the 15th COP to add a new Target in the future framework on "supporting and protecting those individuals and groups who are at the forefront of defending biodiversity and human rights and who are often at risk because as a result" is both relevant and essential (CBD/WG2020/2/4, p. 48).

With regard to the role of principles and approaches in the interpretation of the Framework Convention and biodiversity-related instruments in the post-2020 perspective, the Zero Draft has emphasized the principle of integration of the requirements to protect biodiversity across the spectrum of policy-making. Thus, among the tools and solutions for implementation and mainstreaming of new 2030 Action Targets, the Zero Draft includes

> Integrat[ing] biodiversity values into national and local planning, development processes, poverty reduction strategies and accounts, ensuring by 2030 that biodiversity values are mainstreamed across all sectors and that biodiversity-inclusive strategic environmental assessments and environmental impact assessments are comprehensively applied. (CBD/WG2020/2/3, p. 9).

Principles have also served to limit the way States Parties might adopt exceptions and derogations in biodiversity-related conventions—even in the case of the CBD, whose Article 37 prohibits reservations. Derogations and exceptions to the rules in biodiversity-related conventions allow the participation of a wide range of States with different approaches, national interests and circumstances, which would otherwise not have subscribed to a treaty. In practice, however, derogations and exceptions lead to a fragmented legal scenario in which the strict application of the rules coexists with a more flexible and lenient approach, meaning that certain practices may be legal in some countries and not in others. For example, the circumstances of each State Party to the Bern Convention on the Conservation of European Wildlife and Natural Habitats may justify the different measures that each signatory took to comply with them, but these divergences are detrimental to the convention's original objectives and to the principle of conservation. The Bern Convention has achieved a high level of convergence of national laws, but has been less successful in terms of compliance and enforcement. For the EU and its Member States, which are both parties to the Bern Convention, the conditions for derogations under the Habitats Directive reinforce the conditions to be met under the convention's articles 8, 9 and 16. Furthermore, the interpretation adopted by the European Court of Justice on the Habitats Directive has limited the scope of the exceptions and derogations provided for in some of the convention's most important provisions, creating a legal system in which the causes of limited compliance can be identified, examined and remedied through the application of appropriate principles and approaches.

This is possible because principles and approaches are also points of reference in dispute settlements between States as well as in domestic courts with regard to the implementation of national laws transposing international instruments. Despite their contested legal nature, they can help to interpret general and vague rules that require further elaboration. Thus, in settlements between States, the International Court of

Justice (ICJ) has applied the preventive and precautionary principles and procedural principles related to the management of the components of nature and biodiversity in the Gabcikovo Nagymaros case (Hungary v. Slovakia), Pulp Mills on the River Uruguay (Argentina v. Uruguay) as well as in the more recent cases of Certain Activities Carried Out by Nicaragua in the Border Area (Costa Rica v. Nicaragua), Compensation Owed by the Republic of Nicaragua to the Republic of Costa Rica, in which the ICJ awarded compensation for the impairment or loss of environmental goods and services of the area in question and also for measures taken to restore the wetland—thus taking a step forward in recognizing the due protection of the principle of conservation and ecosystem services. Regarding their application by domestic courts, principles that inspire management of areas of protection not only serve as a limit to State sovereignty but also offer solutions in domestic conflicts involving the natural environment, economic interests and human rights. Thus, Decision X/2 adopting the Strategy Plan for Biodiversity 2011–2020 and the Aichi Targets foresaw that "where trade-offs between biodiversity protection and other social objectives exist, they [could] often be minimized by using approaches such as spatial planning and efficiency measures". However, the growing conflicts between biodiversity and economic sectors such as agriculture, forest, fisheries, tourism, energy and so on have raised doubts about the results of these approaches. In the case of agriculture, experts' reports and the Court of Auditors of the European Union have expressed their despondency regarding the methodology applied to make it possible (2020). In practice, public authorities grant permits and licences, actively tolerate trespasses, and allow landfilling and waste spilling in national parks and protected seas—thus ignoring the principles of transparency, environmental impact assessment, and the updating of permits and licences in line with developments in scientific knowledge. Recently, the Court of Justice of the European Union (CJEU) has adopted important criteria to limit the discretionary powers of Member States regarding the adoption of environmental law reforms to protect economic interests at the expense of biodiversity. As pointed out by Squintani, the recent EU case law in the preliminary rulings presented by the national judges regarding the Habitats Directive shows the importance of further clarification of the standards to pursue a programmatic approach in nature conservation and to evaluate science-based evidence in court (Squintani 2020, p. 129).

One of the results of the conflicts between human rights and biodiversity concerns in the processes of implementation of biodiversity-related conventions has been the displacement of indigenous peoples. Biodiversity projects and actions, despite being scrutinized by national and international authorities from the perspective of the principles enshrined in the UN Declaration on indigenous peoples as well as the CBD, have violated human rights in several countries in Africa and Asia. To prevent and avoid bad practices, NGOs have adopted principles for their action that have become reference points for national institutions in need of best practices and specialization in a legal regime which, in the case of developing countries, has been imposed on their national legal systems from above.

NGOs such as WWF have adopted their own principles that further develop the ones enshrined in United Nations instruments in an attempt to adapt them to practices on the ground in which they are key actors in the implementation of international rules and projects (Larsen and Springer 2008). However, NGOs have also gathered information on cases that highlight the shortcomings and loopholes of international instruments and policies whose implementation has led to the violation of the basic human rights of indigenous peoples.[5]

The principle of sustainable development was incorporated in the CBD both as a goal and as a principle regarding sustainable use. It was developed further in binding and non-binding instruments adopted by the CoP and governing bodies. Its importance as a principle of international law and international environmental law has been widely examined (Rodrigo 2015). Sands considers it to be a concept that breeds other principles:

> The need to preserve natural resources for the benefit of future generations (the principle of inter-generational equity);
>
> The aim of exploiting natural resources in a manner, which is sustainable or prudent or rational or wise or appropriate (the principles of sustainable use);
>
> The equitable use of natural resources, which implies that use by one state must take account of the needs of other states (the principle of equitable use or intra-generational equity);
>
> The need to ensure that environmental considerations are integrated into economic and other development plans, programmes and projects, and that development needs are taken into account in applying environmental objectives (the principle of integration) (Sands et al. 2018, p. 219).

Nonetheless, from the post-2020 perspective, attention should focus on the ability of the principle of sustainable development to bring together different legal regimes and scenarios of biodiversity, from the global perspective of the sustainable development goals (SDGs) to the sectorial approaches of biodiversity-related conventions. According to Lowe, this principle plays the role of "a meta-principle, acting upon other legal rules and principles – a legal concept exercising a kind of interstitial normativity, pushing and pulling the boundaries of true primary norms when they threaten to overlap or conflict with each other" (Lowe 1999, p. 31). While attention has so far concentrated on potential conflicts (Van Asselt 2012), the future will undoubtedly focus on cooperation between regimes and agendas (see Chap. 1 in this book), as the Zero Draft recognizes that the

> 2030 Agenda for Sustainable Development, a global plan of action for people, planet and prosperity will be implemented at the same time as the framework and will create opportunities for synergies but, also, the need to account for and balance trade-offs (Zero Draft, CBD/WG2020/2/3, p. 3).

---

[5]As an unintended result of the implementation of the Aichi Targets on the creation of areas of protection, indigenous peoples have been forcibly displaced from their lands during the work to create national parks, conservation areas and private hunting concessions by both States and NGOs in countries such as Cameroon, Congo, and the Central African Republic (Clarke 2019; OECD 2016).

Furthermore, the principle of sustainable development remains the guiding principle for policy changes at global and national levels and was invoked to this end at the United Nations Summit on Biological Diversity in September 2020. The theme chosen for the plenary session of this Summit was the "Urgent Action on Biodiversity for Sustainable Development"[6] and the subheading of the Leaders' Pledge for Nature was "United to Reverse Biodiversity Loss by 2030 for Sustainable Development". The purpose of this session was to acknowledge the contribution of biodiversity to sustainable development, and to renew the CBD goals involving states and all sectors of society with "whole-of-government and whole-of-society approaches and investments to restore, sustainably use, conserve and equitably share benefits from biodiversity" (GA 2020a, p. 3). However, far from applying an ecocentric or a biocentric approach, the anthropocentric approach that informs sustainable use remains the priority. Biodiversity should be saved through "the green recovery of economies" and biodiversity loss is an economic risk "identified by business leaders as one of the top risks to global business" (GA 2020a, p.4). Therefore, in order to save biodiversity it is first necessary to save the world economy. The principle of sustainable use will be further examined as a specific principle of biodiversity-related conventions.

2.4 Specific Principles of the Framework Convention on Biological Diversity and the Binding and Non-Binding Instruments That Develop It

Beyond the preamble and Article 3 of the CBD, with its particular codification of general principles, many provisions of the convention and its protocols include direct and indirect references to principles and approaches in varying degrees of legal consolidation. Other supplementary principles and approaches, which are not expressly formulated in the framework convention but are subsequently developed by CoPs and governing bodies, are instrumental to the achievement of the convention's objectives and purposes and, as such, are the legal expression of the mandates and powers of delegation vested in the CBD institutions—in particular, the power to update the treaties in accordance with the development of scientific and technical knowledge and the best possible practices adopted by States Parties. Moreover, another upshot of the intrinsic relationship between the biodiversity-related conventions and scientific reports is the fact that concepts widely used not only by legal instruments but also by reference scientific reports are finally elevated to legal status in order to express the expectations regarding the behaviour of States: for example, the principle of restoration, the ecosystem approach, and the principles and

[6]The Biodiversity Summit was convened in response to CBD Conference of the Parties Decision 14/34 and UNGA Resolutions 73/234 and 74/269, which determined the scope, modalities, format and organization of the Summit.

approaches related to the management of areas of protection such as the interconnectivity principle. They all apply to the objectives of the CBD which are in themselves principles and approaches underpinning the interpretation and implementation of the biodiversity legal framework, as well as the design of policies that should incorporate biodiversity requirements, as declared in the 2011–2020 Strategic Plan and the Aichi Targets which set the following mission:

> To take effective and urgent action to halt the loss of biodiversity in order to ensure that by 2020 ecosystems are resilient and continue to provide essential services, there by securing the planet's variety of life, and contributing to human well-being, and poverty eradication. To ensure this, pressures on biodiversity are reduced, ecosystems are restored, biological resources are sustainably used and benefits arising out of utilization of genetic resources are shared in a fair and equitable manner; adequate financial resources are provided, capacities are enhanced, biodiversity issues and values mainstreamed, appropriate policies are effectively implemented, and decision-making is based on sound science and the precautionary approach. (CBD COP 10 Decision X/2).

Therefore, the principle of conservation is a goal and a principle of biodiversity-related conventions which has evolved in line with scientific knowledge and has led to other specific management principles such as the ecosystem approach and the restoration approach which are also interconnected with other biodiversity-related conventions such as the Ramsar Convention. As Koester says, the concept of conservation "is rather weakly defined, but the vagueness of the concept entails opportunities to understand it in the light of contemporary needs" (Koester 2018, pp. 543–544). Conversely, vagueness and ambiguity at the service of national interest have all too often undermined the goals and purposes of biodiversity-related conventions. The soft law nature of the instruments chosen to develop principles and approaches, such as the *Addis Ababa Principles and Guidelines on sustainability on the use of biodiversity*, shows that the COP and governing bodies recognize the limits of consensus-adopting principles and best practices that take into account the differences between the various regional blocs of States in order to enhance the implementation of commitments and promote capacity-building, and at the same time seek to limit the leeway States have in the interpretation of the dispositions of the CBD and biodiversity-related conventions. In addition, the Addis Ababa Principles and Guidelines aim to enable policy-makers, individuals and enterprises to adopt standards of behaviour at the national level, since they "provide a framework for assisting governments, indigenous and local communities, resource managers, the private sector and other stakeholders, about how to ensure that their uses of biological diversity will not lead to its long-term decline" (CBD Secretariat 2004, p. 3). Upgrading these practical principles by incorporating them into a convention or a new protocol to the CBD would have been the ideal measure, but may not have been acceptable to all States Parties to the CBD.

The ecosystem approach has grown steadily in importance ever since its incorporation in the CBD as a soft law operative guidance linked to the conservation and protection of ecosystems, natural habitats, and species. It is now one of the principles embodied in many of the biodiversity-related conventions, and works in combination with the other principles and approaches linked to conservation objectives, such as the precautionary approach, the principle of sustainable use and the principle of integration. This principle and the ecosystem-based adaptation approach are also called upon to play a fundamental role in combining the fight against climate change with the protection of biodiversity. Thus, the ecosystem-based adaptation approach will "contribute to climate change *mitigation* by reducing the emissions that transpire from habitat loss and ecosystem degradation" (UNEP 2019; CBD 2016).

The principle of sustainable use is "[a]n overarching principle" that has been widely adopted in biodiversity-related treaties and in MEAs in general, as examined in several articles in this book. Sustainable use is defined as "the use of components of biological diversity in a way and at a rate that does not lead to the long-term decline of biological diversity, thereby maintaining its potential to meet the needs and aspirations of present and future generations" (Art.2 CBD). In the case of the Ramsar Convention on wetlands of international interest, the use of one of the components of biodiversity requires a "wise use concept" which entails that "parties are expected to maintain the ecological character of the wetlands on their territories", as examined by Gardner in this book. The scope of this open concept has been progressively restricted by criteria adopted in the framework of the CBD and biodiversity-related conventions bodies and committees or by the case law of the EUCJ (Squintani 2020).

When adopting the Sustainable Development Goals in 2015, States undertook to "protect, restore and promote sustainable use of terrestrial ecosystems, sustainably manage forests, combat desertification, and halt and reverse land degradation and halt biodiversity loss". In its Goal 15, the principle of sustainable use is among the suggestions for a revised version of Target 7, as proposed in the Zero Draft. However, its formulation is based on an anthropocentric approach that seeks to accommodate human needs for the sustainable use of wildlife and fails to take account of the measures to contain consumption that the effects of the pandemic have shown to be essential. The new version of Target 7 seeks to:

> Enhance the sustainable use of wild species providing, by 2030, benefits, including enhanced nutrition, food security and livelihoods for at least [X million] people, especially for the most vulnerable, and reduce human-wildlife conflict by [X%] (CBD/WG2020/2/4, p. 25).

As Morgera notes, the principle of benefit sharing is best understood as a general principle of international law that has affected diverse areas such as water, fisheries, climate change, land and food (Morgera 2016, p. 332.) But as the third objective of the CBD, it is in the context of the biodiversity-related conventions that this principle has been deployed most effectively and, in particular, in its Nagoya Protocol on access to genetic resources and benefit sharing. For Morgera, [it] has also emerged under the CBD as a component of the ecosystem approach, in conjunction with the benefit-sharing obligation arising from the use of indigenous peoples' and local communities' traditional knowledge (Morgera 2016, p. 326).

Despite the widespread recognition of this principle, world leaders at the September 2020 Biodiversity Summit acknowledged that "progress towards the fair and equitable sharing of benefits resulting from access to genetic resources has been limited, and on-going advances in biotechnology and synthetic biology alongside the increased use of genetic sequence data currently pose challenges for benefit sharing." Nevertheless, the enormous potential economic benefits also encouraged the leaders to claim that "a multilaterally agreed solution to this challenge that ensures the fair and equitable sharing of benefits while facilitating access to genetic resources has the potential to unleash substantial benefits across many sectors, including health" (Leaders' Pledge 2020).

The principles relating to the rights of indigenous peoples and local communities are an example of both substantive and procedural functions, as they recognize rights and the necessary guarantees for their enjoyment, in particular when these principles overlap with other sectors. Thus, the attempts to achieve biodiversity goals in areas of special protection have faced criticism for having to choose between the rights of nature and the rights of indigenous people. Making these interests mutually supportive requires respecting the principle of Free, Prior, Informed Consent (FPIC) of indigenous peoples as enshrined in the CBD as well as in international instruments such as the United Nations Declaration on the Rights of Indigenous Peoples, and Convention 169 of the International Labour Organization (ILO). These instruments also recognize the indigenous peoples' right of access to the multiple advantages and benefits in the framework of UN initiatives. An example is REDD+, whose working group on Operational Guidelines for Obtaining FPIC has adopted four principles for the implementation of its projects that are instrumental and operational since they seek an implementation that respects the human rights of indigenous people. On the other hand, the guarantee and enjoyment of these rights depend on the observance of basic rights of political participation and property rights, as well as good practices in applying the principle of environmental impact assessment in which their opinions are heard and duly taken into account. The difficulties regarding the implementation of the Escazu Agreement and its rejection by Chile have their origin in the limitations of the legal institution of indigenous consultation and the case law of the Supreme Court which, following the example of the CBD, does not consider the environmental impact assessment to be binding even when indigenous peoples' territories are affected by infrastructure works that are justified by the general interest but destroy or seriously affect biodiversity.

2.5 Principles for the Post-2020 Scenario According to the Summit on Biodiversity and the Zero Drafts of the CBD COP 15

The three principles highlighted in the preliminary documents of the Summit on Biological Diversity and finally enshrined in the Leaders' Pledge for Nature remain the same as the ones derived from the objectives of the CBD: conservation, sustainable use, and equitable sharing of benefits of biological diversity. These are now also the principles underlying the Agenda for Sustainable Development 2030 (GA 2020a,

p. 3). However, in 2010, as recalled in the process of preparation of the future CBD CoP 15, the Parties to the CBD committed themselves to the same goals and to "take effective and urgent action to halt the loss of biodiversity" through the Strategic Plan for Biodiversity 2011–2020 and its Aichi Targets. In these past 10 years, national action plans and strategies adopted by States Parties have not achieved the goals pursued due to the lack of ambition and financial resources. What needs to change now, in order to avoid reaping the same harvest of failure in the coming years, is a global commitment to transnational cooperation and the provision of financial support to countries in need of technology and resource transfer to enable them to implement their national commitments more ambitiously. In their constant strategy of supporting green multilateralism, the EU and its Member States have already committed themselves to support transnational cooperation based on "long-lasting engagement . . . in conserving and restoring the ecological connectivity and integrity of ecosystems to support the natural movements of animals, necessary for their survival and well-being" (EU 2020, p. 3). With this pledge, the EU also subscribes to principles that will be key in the future development of both soft law and binding instruments: the principle of restoration to which the UN has dedicated an entire decade, and the principles of ecological connectivity and integrity of ecosystems. These state-of-the-art principles for the conservation of biological diversity in the times of the coronavirus must be closely connected with the "One Health" approach that was already adopted by the CBD bodies some years ago and has now become a central player in the fight against the coronavirus, when the loss of biodiversity and the changes in land use may be the causes of the zoonosis.

Last but not least, the CBD COP 15 needs to address the biodiversity governance gap (Silver et al. 2020), because international environmental law does not address biodiversity monitoring and conservation seriously. The endemic problem of the lack of enforcement of, and compliance with, international environmental law has its own expression in the biodiversity-related conventions. In order to attract the participation of as many States as possible, the obligations laid out in these conventions are ambiguous and flexible, and their clauses contain a list of exceptions and derogations that permit States to implement their own visions of biodiversity conservation and sustainable use. This has led to a fragmented legal scenario that undermines the conventions' objectives. Nevertheless, in the Leaders' Pledge, States committed to a "transformational post-2020 global biodiversity framework for adoption at the CBD CoP 15" that will be "backed up by a strong monitoring and review mechanism, and means of implementation commensurate with the challenge of halting and reversing the decline in biodiversity". If this commitment is fulfilled, it will certainly make a difference.

2.6 Conclusion

In the post-2020 scenario, the hope is that the outcome of the CBD CoP 15, to be held in Kunming in 2021, will provide the formulas and solutions that humankind needs to reconcile itself with nature, to survive and to place biodiversity "at the heart of sustainable development" (GA 2020b). Agreement is needed on the principles and approaches that will allow States to fight pandemics and "the continued loss of biodiversity and the threat that this poses to human well-being and the prospects for achieving the 2030 Agenda for Sustainable Development and for reaching the three objectives of the Convention" (Zero Draft p. 1). These principles should give States, international organizations and actors such as indigenous peoples and local communities the tools needed to adapt to risks and uncertainties in different timeframes and geographical scenarios, respecting the imperative of scientific knowledge. These principles and approaches will make possible the acceptance of future measures not as conventional obligations but as voluntary commitments that allow States to accommodate their international commitments to their national interests and capacities, because they would not accept it under any other formula that would compromise their sovereignty. Thus, soft instruments have been adopted by States Parties in the CoP as well as by committees in order to further develop systemic, substantive and operative principles able to implement the general rules contained in the CBD or other related-biodiversity conventions. These principles help to structure a constantly evolving legal system that adapts to new challenges as well as old ones that have not been successfully addressed. Before a consensus is reached, principles serve to create the legal horizon required to face uncertainty and risk and other outstanding issues that cannot be neglected for much longer. The failures recorded to date must be used as a basis for finding solutions rather than serving as excuses. These are the high stakes for which the CBD CoP 15 must play in Kunming in 2021.

References

Boisson de Chazournes, L., & Maljean du Bois, S. (2016). Princlpes de Droit International de l'Environnement. *Jurisclasseur Environnement et Development Durable, Fasc. 2010*, 1–22.

Campins Eritja, M. (2015). De Kioto a París: ¿Evolución o Involución de las Negociaciones Internacionales sobre el Cambio Climático? *Instituto Español de Estudios Estratégicos, 15*, 1–19.

CBD. (2016). *Synthesis Report on Experiences with Ecosystem-Based Approaches to Climate Change Adaptation and Disaster Risk Reduction* (CBD Technical Series No. 85).

CBD. (2020, February 29). *Report of the Open-Ended Working Group on the Post-2020 Global Biodiversity Framework on its Second Meeting* (CBD/WG2020/2/4).

CBD Secretariat. (2004). *Addis Ababa principles and guidelines for the sustainable use of biodiversity*. Montreal: Secretariat of the Convention on Biological Diversity.

Clarke, C. (2019). *In and around Cameroon's protected areas: A rights-based analysis of access and resource use agreements between indigenous peoples and the state*. Moreton-in-Marsh: Forest Peoples Programme.

EU. (2020, March 10). Statements by the EU and its Member States, 13th meeting of the Conference of the Parties to the Convention on the Conservation of Migratory Species of Wild Animals (CMS COP 13) (Gandhinagar, India, 15–22 February 2020), 6320/20.

Fajardo, T. (2018). Environmental law principles and general principles of international law. In L. Krämer & E. Orlando (Eds.), *Principles of environmental law, Elgar Encyclopedia of environmental law* (Vol. VI, pp. 38–51). Cheltenham: Eduard Elgar Publishing.

GA. (2019). Note by the President, Revised draft text of an agreement under the United Nations Convention on the Law of the Sea on the conservation and sustainable use of marine biological diversity of areas beyond national jurisdiction, 18 November 2019, *A/CONF.232/2020/3*.

GA. (2020a). *Concept Note of the President on the Summit on Biodiversity*, 24 July 2020.

GA. (2020b). *Leaders' Pledge for Nature United to Reverse Biodiversity Loss by 2030 for Sustainable Development*.

ICJ. (2015). Certain Activities Carried Out by Nicaragua in the Border Area (Costa Rica v. Nicaragua) and Construction of a Road in Costa Rica along the San Juan River (Nicaragua v. Costa Rica), Judgment, I.C.J. Reports 2015, p. 665. Retrieved December 16, 2020, from https://www.icj-cij.org/public/files/case-related/150/150-20151216-JUD-01-00-EN.pdf.

Juste Ruiz, J. (1997). *Derecho Internacional del Medio ambiente*. Madrid: McGraw Hill.

Kiss, A. (1993). Les traités-cadres: une technique juridique caractéristique du droit international de l'environment. *Annuaire Français de Droit International, XXXIX*, 792–797.

Koester, V. (2018). Environmental principles and concepts in biodiversity treaties. In L. Krämer & E. Orlando (Eds.), *Principles of environmental law, Elgar Encyclopedia of environmental law* (Vol. VI, pp. 538–554). Cheltenham: Eduard Elgar Publishing.

Kohen, M., & Schramm, B. (2017). *General principles of law*. Oxford Bibliographies.

Krämer, L. (2018). Introduction to Volume VI. In L. Krämer & E. Orlando (Eds.), *Principles of environmental law, Elgar Encyclopedia of environmental law* (Vol. VI, pp. 1–10). Cheltenham: Eduard Elgar Publishing.

Leal-Arcas, R. (2011). Top-down versus bottom-up approaches for climate change negotiations: An analysis. *The IUP Journal of Governance and Public Policy, 6*, 7–52.

Lowe, V. (1999). Sustainable development and unsustainable arguments. In A. Boyle & D. Freestone (Eds.), *International law and sustainable development: Past achievements and future challenges* (pp. 19–37). Oxford: Oxford University Press.

Morgera, E. (2016). The need for an international legal concept of fair and equitable benefit sharing. *European Journal of International Law, 2*(2), 353–383.

Morguera, E. (2018). Fair and equitable benefit-sharing. In L. Krämer & E. Orlando (Eds.), *Principles of environmental law, Elgar Encyclopedia of environmental law* (Vol. VI, pp. 323–350). Cheltenham: Eduard Elgar Publishing.

OECD. (2016). *OECD Wacht Survival International vs. WWF*. Retrieved December 1, 2020, from https://complaints.oecdwatch.org/cases-es/Case_457

Rodrigo, A. J. (2015). *El Desafío del Desarrollo Sostenible. Los principios de Derecho internacional relativos al desarrollo sostenible*. Madrid: Marcial Pons.

Sands, P., Peel, J., Fabra, A., & Mackenzie, R. (2018). *Principles of international environmental law* (4th ed.). Cambridge University Press.

Sliver, J., Acton, L., Campbell, L., & Gray, N. (2020). How a global ocean treaty could protect biodiversity in the high seas. *The Conversion*. Retrieved December 1, 2020, from https://theconversation.com/how-a-global-ocean-treaty-could-protect-biodiversity-in-the-high-seas-139552.

Squintani, L. (2020). Balancing nature and economic interests in the European Union: On the concept of mitigation under the Habitats Directive. *Review of European Comparative & International Environmental Law, 29*, 129–137.

UNEP. (2019). *Ecosystem-based Adaptation: Briefing Note Series*. Retrieved December 1, 2020, from https://www.unenvironment.org/explore-topics/climate-change/what-we-do/climate-adaptation/ecosystem-based-adaptation

Van Asselt, H. (2012). Managing the fragmentation of international environmental law: Forests at the intersection of the climate and biodiversity regimes. *New York University Journal of International Law and Politics, 44*, 1205–1278.

Willmore, C. (2017). Sovereignty, conservation and sustainable use. In E. Morgera & J. Razzaque (Eds.), *Biodiversity and nature protection law, Elgar Encyclopedia of environmental law* (Vol. III, pp. 31–43). Elgar Publishers.

Chapter 3
Gaps in International Biodiversity Law and Possible Ways Forward

José Juste Ruiz

Contents

3.1 Introduction 36
3.2 Gaps in International Biodiversity Law 37
  3.2.1 Structural Gaps 37
  3.2.2 Regulatory Gaps 39
  3.2.3 Implementation and Enforcement Gaps 41
3.3 Building an Integrated Biodiversity Regime 43
  3.3.1 Integration Between Biodiversity-Related Conventions 43
  3.3.2 Conclusion of New Conventions 46
  3.3.3 Operationalization of Principles 49
3.4 Conclusion 53
References 54

Abstract Building on the report on gaps in international environmental law prepared in 2018 by the UN Secretary-General, this chapter focuses on the existing gaps in international biodiversity law and explores possible ways forward to a more effective and integrated legal regime. The analysis indicates the existence of structural, regulatory and implementation gaps that seriously undermine the efficiency of the regime. The chapter explores the possibility of building a more integrated international legal regime for biodiversity protection by analysing three possible avenues: greater integration between existing conventions, the conclusion of new conventions (such as the impending agreement on marine biodiversity in areas beyond national jurisdiction) and the operationalization of principles proclaiming the collective interest of all States in the conservation and sustainable use of biodiversity. The conclusions suggest that, in order to bridge existing gaps in the

This chapter is an outcome of the research project entitled "El pilar ambiental del desarrollo sostenible: evolución en el marco del Derecho internacional y Europeo" (DER2017-85443P MINECO/AEIFEDER, UE).

J. Juste Ruiz (✉)
Universitat de Valencia, Faculty of Law, Valencia, Spain
e-mail: jose.juste@uv.es

M. Campins Eritja, T. Fajardo del Castillo (eds.), *Biological Diversity and International Law*, https://doi.org/10.1007/978-3-030-72961-5_3

post-2020 framework, Parties to biodiversity conventions should focus not only on policy solutions but also on more committed legal actions to preserve the Earth's biodiversity.

3.1 Introduction

In recent years the discourse on gaps in international environmental law has become fashionable. An important driver of the revitalization of the debate has been the report on gaps in international environmental law prepared in 2018 by the Secretary-General of the United Nations as part of the process "towards a global pact for the environment" that was launched by General Assembly Resolution 72/277 (Juste Ruiz 2020).

The concept of gaps in the international biodiversity regime evokes the twofold problem of material scope and functional efficiency. The first aspect raises the question of completeness: does the biodiversity regime cover all aspects of the subject or are there loopholes remaining to be filled? The second aspect raises the question of efficiency: does the regime regulate matters in a fully efficient way or are there challenges yet to be addressed? The problem of gaps is further complicated by the oversized scope of the concept of biodiversity, a term that is clearly over-inclusive. The inherent complexity of the issue is also affected by the impact of a North-South divide that distorts the contours of the debate (Jeffrey et al. 2008).

The issue examined is far from being merely theoretical. The importance and value of biodiversity and ecosystem services and their progressive decline world-wide are well-established scientific facts. Biodiversity loss has been identified as one of the nine planetary boundaries that determine the self-regulating capacity of the Earth system in the Anthropocene (Rockström et al. 2009). The 2018 report from the Secretary-General on gaps in international environmental law clearly summarizes the gravity of the problem and justifies the need to address it:

> There is scientific consensus that, globally, biodiversity is being lost at an alarming rate. Threats to biodiversity come from a multitude of direct and indirect sources and activities, ranging from habitat fragmentation, pollution and the introduction of alien invasive species to climate change. Drivers of biodiversity loss are often complex, multiple and interlinked, and require the interplay of many different instruments. Many of the threats, as well as the habitats, ecosystems or species to which they apply, do not respect national boundaries or are found in areas beyond national jurisdiction. At the same time, in the light of the complexity of the issue, the science is incomplete or lacking in some aspects. (para. 40)

This chapter takes the broad approach to the concept of gaps that guides the Secretary-General's report encompassing both the lacunae in international biodiversity law and its inefficiencies in solving the challenges posed by the existing regime. The following sections will examine, first, the main types of gaps that exist in the international biodiversity regime and, then, the possible ways to overcome such gaps.

3.2 Gaps in International Biodiversity Law

The gaps in international biodiversity law are of three main kinds. They affect the structure of the biodiversity regime (structural gaps), the regime's ability to comprehensively regulate and govern the conservation and sustainable use of biodiversity (regulatory and governance gaps) and the effective implementation of biodiversity law at the international and national levels (implementation gaps).[1]

3.2.1 *Structural Gaps*

The international legal regime for the conservation of biodiversity comprises a suite of multilateral environmental agreements (MEAs) that cover a wide range of different but interdependent matters.

The conventions that make up the regime are so different that classifying and arranging them into a coherent structure poses something of a challenge. For descriptive purposes, the existing global conventions may be grouped ideally into four concentric circles. The innermost circle contains the core general biodiversity conventions, namely the Convention on Biological Diversity (1992), the Cartagena Protocol on Biosafety (2000), the latter's Nagoya–Kuala Lumpur Supplementary Protocol on Liability and Redress (2010) and the Nagoya Protocol on Access to Genetic Resources and Benefit Sharing (2010). Moving outwards, the second circle is composed of the five independent conventions that address specific biodiversity matters, namely the International Whaling Convention (1949), the Wetlands Convention (1971), the World Heritage Convention (1972), the Convention on International Trade in Endangered Species of Wild Fauna and Flora (1973) and the Bonn Convention on Migratory Species (1979). The third circle includes the conventions dealing with separate but closely related issues, such as the fishing and conservation of marine living resources (UNCLOS 1982, the Fish Stocks Agreement 1995, FAO conventions on fisheries), terrestrial resources (Desertification Convention 1995), and FAO conventions on food and agriculture (the 1951 IPPC and 2001 ITPGRFA). The outermost circle is made up of conventions pertaining to separate fields such as health, trade, investment, intellectual property and human rights that have cross-cutting relationships with the international regime on biodiversity. There are also a great many regional conventions that address different aspects of biodiversity conservation and sustainable use whose harmonization with the global biodiversity conventions also raises a number of difficult challenges.

The international legal regime on biodiversity is characterized by fragmentation. The disparate suite of biodiversity-related conventions forms a piecemeal legal

[1]Most of these gaps could be considered as being of a normative nature but in the current atmosphere of multilateral environmental negotiations the existence of normative gaps is politically "unmentionable" (French and Kotzé 2019).

structure that falls short of laying out a fully integrated regime. Having been developed in a disconnected, *ad hoc* manner, the biodiversity conventions have distinct objects and legal structures and utilize different regulatory techniques. The Convention on Biological Diversity (CBD) takes a more holistic approach; it addresses both the direct and indirect causes of biodiversity loss and seeks to mainstream biodiversity considerations into all relevant policy areas. The CBD surely plays a central role in the system, but it cannot be viewed as a genuine framework convention that has authority over the other conventions with which it coexists (Sands & Peel 2012). This clearly follows from the wording of article 22 of the CBD concerning its relationship with other international conventions:

1. The provisions of this Convention shall not affect the rights and obligations of any Contracting Party deriving from any existing international agreement, except where the exercise of those rights and obligations would cause a serious damage or threat to biological diversity.
2. Contracting Parties shall implement this Convention with respect to the marine environment consistently with the rights and obligations of States under the law of the sea.

Under the principles of international law, each biodiversity MEA represents an independent legal instrument "with different scopes, subject matter, type of norms and methodological approach" (Jóhannsdóttir et al. 2010, p. 142). Preserving the legal autonomy of each convention regime is an important objective for States, especially for the major powers that find this disparate structure well suited to their particular interests. The operationalization of the legal autonomy of each biodiversity convention is further dependent on the congruence of their respective memberships; significant disparities in constituencies might restrict the possibility of legal interaction and cooperation between related treaties. Ultimately, the exacerbation of "institutional sovereignty" under the biodiversity-related treaty regimes may lead to episodes of environmental "problem shifting", rather than problem-solving, across multiple institutional and environmental boundaries. Prime examples of the challenge can be seen in the impact of afforestation and reforestation projects on biodiversity loss (Kim and Bosselmann 2013) and the possible impact of ocean fertilization and geoengineering activities on marine biodiversity (Juste Ruiz 2018).

At present, the amalgamation of different biodiversity-related conventions does not provide a fully efficient framework for biodiversity regulation and governance. The system lacks both an inherent hierarchy and a typical centre. The fragmentation of the legal regime generates functional overlapping and a duplication of effort and it may be a source of legal gaps. As stated on the Secretary-General's 2018 report on gaps in international environmental law:

> The legal instruments for the conservation of biodiversity have developed without an overall strategy and have no coherent structure. This situation leaves some issues without specific, legally binding regulation, including on the conservation and sustainable use of forests, the pollution of marine areas by land-based plastic waste, the protection of soil, the use of pesticides, noise pollution, human rights and biodiversity, the Arctic area, nanomaterials and some geo-engineering processes. (para. 41)

The report adds that investment and intellectual property instruments have not evolved to include concerns on biodiversity (para. 47, 48). However, the report's list of structural gaps in the international biodiversity regime is not exhaustive; issues such as land and soil protection, the proliferation of alien invasive species and issues relating to liability and redress for damages to biological diversity—other than for biosafety-related activities—also lack comprehensive regulation.

3.2.2 Regulatory Gaps

The conventions relating to biodiversity use for their implementation a variety of regulatory techniques that range from soft-law approaches to more stringent regulations.

The CBD adopts a predominantly soft-law approach whereby the established commitments are to be performed "as far as possible and as appropriate". Under this approach, the convention sets strategies and objectives that must be implemented through national action plans established under the domestic law of each Party. For example, in 2002 the CBD CoP established biodiversity targets for 2010, which were followed by the Strategic Plan for Biodiversity 2011–2020 and its Aichi Biodiversity Targets. As prescribed in article 6 of the CBD, these strategic objectives and the corresponding measures adopted for their implementation are to be integrated into national strategies and action plans or programmes in accordance with each Party's "particular conditions and capabilities".

Other biodiversity conventions require the Parties to adopt more specific and direct regulatory measures. Different regulatory tools, based on different criteria, are applied for the implementation of each convention's respective objectives. Typically, these tools consist of establishing lists of protected species (IWC), habitats (CMS) or sites (WHC), regulating permits for trade on protected species (CITES), and, in some cases, establishing protected areas (CBD). The 1979 Bonn Convention (CMS) provides that Parties that are Range States of migratory species listed in Annex I must endeavour to conserve and restore habitats and to prevent or minimize the adverse effects of activities that seriously impede or prevent migration. For species that have an "unfavourable conservation status", the convention requires Range States to conclude agreements to benefit those species. So far, seven regional legally binding agreements have been adopted.[2] In the same vein, CITES stringently regulates trade in approximately 36,000 species that are included in the three

[2]The seven agreements are: Agreement on the Conservation of Seals in the Wadden Sea; Agreement on the Conservation of Populations of European Bats; Agreement on the Conservation of Small Cetaceans of the Baltic, North East Atlantic, Irish and North Seas; Agreement on the Conservation of African–Eurasian Migratory Waterbirds; Agreement on the Conservation of Cetaceans of the Black Sea, Mediterranean Sea and Contiguous Atlantic Area; Agreement on the Conservation of Albatrosses and Petrels; and Agreement on the Conservation of Gorillas and Their Habitats.

appendices to the convention. However, the non-binding nature of the resolutions of the CoPs and the lack of international enforcement measures undermine the effectivity of these regulatory regimes.

One important regulatory feature of the international biodiversity regime is the incorporation of the "ecosystem approach" in article 3 of the Biodiversity Convention and its continued endorsement by the CoP as a paramount governance tool. However, a revision of the CoP decisions on the issue shows that there remain considerable doubts about how to move forward on the ecosystem approach to biodiversity governance, "especially if considered in terms of social-ecological resilience thinking" (Hey 2016, p. 47). Among the various tools employed to achieve the ecosystem objectives of the biodiversity conventions, the establishment of area-based management schemes, such as protected land and/or marine areas, is growing in preference. Article 8 of the CBD states that "each Contracting Party shall, as far as possible and as appropriate, establish a system of protected areas or areas where special measures need to be taken to conserve biological diversity." The establishment of specially protected areas is surely the most comprehensive way to protect biodiversity in the areas concerned, but the efficiency of the system is weakened by the fact that their establishment is not mandatory. The biodiversity conventions invite Parties to establish specially protected areas but do not impose a legal obligation to do so. In addition, in cases of protected areas extending beyond national jurisdiction (like the high seas), there are challenges to ensuring implementation by third States (Scovazzi 2004, 2014).[3] The drawbacks that may result from this non-binding approach are highlighted in the Secretary-General's report on gaps in international environmental law as follows:

> The establishment of protected areas is a primary tool to implement in situ conservation, though its effectiveness is a concern. While target 11 of the Aichi Biodiversity Targets provides for protected areas, it does not impose legal obligations. Instead, parties are urged to develop national and regional plans with a view to contributing to efforts to reach the global Aichi Targets. This approach may not be sufficient to address the interconnectedness of ecosystems activities or uses that take place outside the protected area. (para. 42)

In summary, the biodiversity conventions evince regulatory deficits arising from their predominantly soft-law approach and the "weak or unclear" legal status of the decisions of their respective CoPs (Jóhannsdóttir et al. 2010, p. 143). The legally binding force of the decisions adopted by the governing bodies of the conventions remains unclear, a circumstance that seriously challenges the regulatory capacity of the regime.

[3]In order to protect marine biodiversity, some specially protected areas extending fully or partially into the high seas have been established under regional marine conventions: 14 under OSPAR and one under the Barcelona Convention (the 1999 French–Italian–Monegasque Pelagos sanctuary for marine mammals). In 2010, a marine protected area extending into waters beyond national jurisdictions was established by the UK in the Chagos Archipelago (Dunne et al. 2014; Juste Ruiz 2016).

3.2.3 Implementation and Enforcement Gaps

The gaps in international biodiversity conventions are particularly evident in their surveillance and control mechanisms. Following a pattern common to multilateral environmental agreements, the main monitoring and verification procedures are reporting and compliance mechanisms. In some cases, the biodiversity conventions provide for more stringent enforcement measures such as inspections and trade restrictions.

Reporting Mechanisms As a rule, States Parties to biodiversity conventions are obliged to monitor the national implementation of the conventions and to comply with the established reporting obligations. In some cases, the obligation of reporting is established in the convention (i.e., WHC, article 29) but in other cases this obligation results from a recommendation of the CoP (i.e., Ramsar, Recommendation 2.1). The conventions on marine living resources provide for specialized systems of reporting adapted to the specific activity regulated (Edeson 2016). The effectiveness of the reporting mechanisms depends on a series of factors, most significantly the actual international commitments that are scrutinized (e.g. strict quantitative catch quotas under RFMOs or diffuse, qualitative obligations under the CBD), the administrative and financial capacity of the Parties (especially among developing countries) and the degree of institutionalization of each regime to deal with reporting and monitoring (Cardesa-Salzmann 2017).

The record of reporting in biodiversity-related conventions is poor, but it is slightly improving. While many Parties take the reporting obligations seriously, they do not all do so and even fewer use the reports as a tool to assist in better implementation. One recurrent complaint of States is the sheer amount and overlapping nature of reporting requirements under biodiversity-related conventions (Kiss 2006, pp. 244–245). As we will see later, the harmonization of reporting requirements is currently on the agenda of several biodiversity conventions.

Compliance Procedures In addition to reporting, some global conventions on biodiversity establish different sorts of compliance mechanisms to verify adherence to the convention's obligations and help the Contracting Parties to achieve compliance. The Ramsar Convention, the WHC, CITES, the Cartagena and Nagoya Protocols, and the International Treaty on Plant Genetic Resources for Food and Agriculture (ITPGRFA) have established compliance procedures with different characteristics as to their scope, the institutional bodies involved and response measures. However, several global biodiversity conventions, including the IWC, the CMS, the FSA and the CBD itself, lack a proper non-compliance institutional procedure.

The respective institutional arrangements for compliance mechanisms are fine-tuned to the particular enforcement needs of each biodiversity convention. For instance, CITES has established quite effective compliance processes which give a relevant role to the Secretariat, but without establishing a separate and independent compliance committee. Other biodiversity conventions have established genuine compliance committees, formed by independent elected members, that in some cases exhibit innovative features. For instance, in the mechanism of the 2010 Nagoya Protocol on Genetic Resources the persons appointed to the compliance committee could include representatives of indigenous and local communities (decision NP-1/4, 2014). In a number of important respects, however, the procedures of the compliance committees established by biodiversity conventions do not meet the most advanced standards. Referrals to the committee by the secretariat, by non-State actors or by the committee itself (self-triggering) are generally excluded. Public participation in the meetings of compliance committees is also excluded except in the case of the Nagoya Protocol, where two representatives of indigenous and local communities may participate as observers in the deliberations of the committee (unless the Party concerned requests that the deliberations be closed to observers).

Inspections, Removal from the Lists and Trade Restrictions Some biodiversity conventions, such as the IWC, a few fisheries treaties and the Antarctic conventions provide for inspections by observers appointed by the Parties willing to participate in the scheme on a mutual basis. However, these exceptional schemes "fall well short of independent compulsory inspection" (Birnie et al. 2009, p. 244). Under other biodiversity conventions the governing bodies may adopt measures of a more stringent, punitive character. Such measures can include the removal of deteriorated sites from the lists of the Ramsar Convention or WHC as well as the application of trade restrictions on CITES species. The CITES Conference of the Parties and its Standing Committee, which takes its decisions by a simple majority, have a remarkable track record of issuing recommendations to suspend commercial trade in CITES species with non-compliant parties. However, the enforcement of CITES recommendations on trade suspensions and any inspection and border control of permits must be handled by the competent national authorities. Relying on national enforcement has been identified as a major weakness of CITES

> since enforcement is left to individual states parties, whose domestic wildlife laws, scrutiny and controls vary greatly in scope and stringency of enforcement (Birnie et al. 2009, pp. 689–690)

On the whole, the mechanisms and procedures established for the implementation of the biodiversity conventions have not proven to be fully efficient in ensuring the goals pursued. As concluded in the Secretary-General's 2018 report on gaps in international environmental law, "(i)t is widely recognized that insufficient progress has been made on the implementation" of biodiversity-related conventions:

> The biodiversity cluster of treaties is also characterized by issues of ineffective implementation; ineffectual processes relating to monitoring, reporting, review and verification; and the absence of or inadequate procedures and mechanisms to promote and enforce compliance. (para. 42)

Modern doctrinal approaches suggest an additional method to improve the implementation of the biodiversity conventions consisting in the use of legal indicators, distinct from but complementary to scientific, economic and social indicators, which may help to assess the effectiveness of the Parties' compliance with their biodiversity obligations (Prieur 2018).

3.3 Building an Integrated Biodiversity Regime

Given the insufficient on-the-ground results of the international regime for the conservation of biological diversity, a variety of attempts have been made to improve the existing legal framework in order to build a more integrated regime.

3.3.1 Integration Between Biodiversity-Related Conventions

Against the broader canvas of wider UN institutional reform, UNEP has promoted various initiatives to improve international environmental governance, putting particular emphasis on the need to increase integration among the conventions on biological diversity. Integrating the biodiversity conventions requires identifying their connections and interconnections and establishing cooperative mechanisms to facilitate synergies between them (Chambers 2008).

Cooperation Among Biodiversity Conventions In 1995 the Second Meeting of the Conference of the Parties to the CBD adopted Decision II/13 on "cooperation with other biodiversity-related conventions". The decision called on the Executive Secretary to coordinate with the secretariats of other conventions, with a view to facilitating the exchange of information and experience, exploring the harmonization of reporting requirements, coordinating work programmes and consulting on how other conventions can contribute to the implementation of the CBD. As a follow-up, a web of memoranda of cooperation or understanding was created between the secretariats of CITES, Ramsar, CMS, WHC, ITPGRFA and IPPC. Also, some joint working programmes were established to develop partnerships focusing on common goals and mechanisms to avoid duplication. The Strategic Plan for Biodiversity 2011–2020 and its Aichi Biodiversity Targets provide an agreed-upon overarching framework for cooperation among the biodiversity-related conventions. However, early doctrinal assessments of the work carried out indicate that "little tangible progress has resulted", because increased administrative cooperation has not been translated into better biodiversity outcomes (Jóhannsdóttir et al. 2010, p. 143).

Liaison Groups Between the Conventions In order to increase cooperation, the CBD CoP decided to establish, in August 2001, a Joint Liaison Group (JLG) between the three Rio conventions (CBD, UNFCC and Desertification). Membership in the Rio conventions' JLG is made up of the officers of the conventions' scientific subsidiary bodies, the executive secretaries and members of the secretariats. The JLG's functions are modest: it acts as an informal forum to exchange information, explore opportunities for synergistic activities and increase coordination.

In 2004, CBD CoP Decision VII/26 recognized the role of the Joint Liaison Group and urged "further enhanced cooperation between the Convention on Biological Diversity and all relevant international conventions, organizations and bodies ... to enhance synergies and reduce inefficiencies" (para. 1). The decision requested the Executive Secretary to invite the secretariats of CITES, Ramsar, CMS and World Heritage Convention "to form a liaison group" to enhance coherence and cooperation in their implementation (para. 2). The resulting Liaison Group of Biodiversity-related Conventions (BLG) meets regularly to explore opportunities for synergistic activities and increased coordination and to exchange information. The CBD CoP has repeatedly underlined the important role of the BLG, encouraged more regular meetings, and invited the group to identify options for improved implementation of and cooperation among the biodiversity-related conventions (see, for instance, Decision IX/27 (2008) para. 3 and 4). This recurrent invitation for improved cooperation is an implicit recognition that additional efforts must be made by the BLG in order to enhance the harmonization and effective implementation of the biodiversity-related conventions.

Science–Policy Interface Efforts to increase scientific cooperation started in 2006 when Decision IX/27 of the CBD CoP invited the subsidiary scientific and technical bodies of the three Rio conventions to enhance mutual collaboration in a manner consistent with their respective mandates, governance arrangements and agreed programmes (para. 2, 5, 6). Moving forward to wider integration, CBD CoP Decision X/11 recommended in 2010 that the UN General Assembly consider the establishment of an intergovernmental science–policy platform at the earliest opportunity (para. 2). Following directives of the UN General Assembly and the Governing Council of UNEP, the Intergovernmental Science–Policy Platform on Biodiversity and Ecosystem Services (IPBES) was established by 94 States on 21 April 2012 to act as an independent intergovernmental body.[4] The IPBES's main aims are to strengthen the science–policy interface on issues of biodiversity and ecosystem services for the conservation and sustainable use of biodiversity, long-term human well-being and sustainable development.

Although the IPBES's work is an important contribution to the science–policy interface on biodiversity governance, doctrinal opinions have stressed the difficulties of the IPBES in achieving a better use of scientific knowledge in decision-making (Vadrot 2014). One of the major challenges ahead is the harmonization of the respective policies of the IPBES, an external intergovernmental body, and the Subsidiary Body on Scientific, Technical and Technological Advice, which is a convention body established under article 25 of the CBD.

---

[4]IPBES is not a United Nations body, although UNEP has provided secretariat services to IPBES since 2013.

Simplified Reporting Procedures When dealing with the integration of biodiversity-related conventions, a recurring issue is the need to simplify national reporting procedures and possibly unify them in order to avoid duplication of work.

Building on a preconditions report commissioned by UNEP in 2009, some steps have been taken to ease the reporting procedures for the Contracting Parties to the biodiversity conventions.[5] An initiative to develop joint thematic reporting on inland waters and other wetlands has been adopted under the 5th CBD–Ramsar Joint Work Plan 2011–2020. With the aim of expanding the process of aligning national reporting, assessment and review, in 2018 the CBD CoP adopted Decision 14/27 by which the CoP decided "to commence with synchronized reporting cycles for the Convention, the Cartagena Protocol and the Nagoya Protocol in 2023". The decision invites the respective CoPs to undertake the preparatory measures and calls on the Executive Secretary to assess the cost implications of synchronized reporting cycles, improve and harmonize the user interface and design of national reporting, identify concrete actions to advance synergies in reporting and contribute to the monitoring process for the 2030 Agenda for Sustainable Development. The process of synchronizing reporting cycles, which is limited to the CBD, the Cartagena Protocol and the Nagoya Protocol, is still at an early stage of development.

Holding Joint Meetings of the CoPs Following the example of MEAs in the chemical and wastes sectors,[6] efforts have also been made to improve the clustering of biodiversity conventions by holding joint meetings. In 2014 CBD CoP Decision XII/27 agreed to hold joint meetings of the CoP to the Convention, the Cartagena Protocol on Biosafety (and its Supplementary Protocol on Liability and Redress) and the Nagoya Protocol on Genetic Resources. The decision notes that the process is an iterative one that needs to ensure the availability of financial support for the participation of all legitimate representatives and the streamlining of meeting agendas.

[5]The preconditions report highlights the potential approaches to harmonization of national reporting to the biodiversity-related MEAs and notes the ongoing challenges and difficulties yet to be overcome. The options include a reporting system based on a core report relevant for all treaties involved, supplemented by smaller treaty-specific reports that address the specific information needs of the MEAs involved. In its conclusions, the report states that it is obvious that there is now a need for a more practical approach that addresses the preconditions identified in the report and makes progress towards harmonization.

[6]In 2006, each of the CoPs to the Basel, Rotterdam and Stockholm Conventions agreed to establish an *ad hoc* joint working group on enhancing cooperation and coordination (AHJWG). The AHJWG agreed on a recommendation that included proposals for joint programmatic and administrative activities. Common to all three conventions are several issues that aim to strengthen particular tasks such as the harmonization of national reporting, joint capacity-building activities and public awareness activities, and the organization of an extraordinary meeting of the CoPs to the three conventions. The process has been described as an innovative "bottom-up" approach; it is furthermore country-driven and engages the Parties to the MEAs in the design and follow-up of synergistic efforts. However, the biodiversity MEAs—perhaps because some of them date back 40 years and have been adopted in a piecemeal fashion, often with a lack of effective coordination and engagement at the national level—have failed to deliver such concrete and synergistic results.

In 2018, Decision 14/32 on the review of the experience in holding concurrently meetings of the CoP to the three legal instruments notes with satisfaction that the concurrent meetings have allowed for increased integration among the convention and its protocols. However, the decision also notes that further improvements in the functioning of the concurrent meetings are desirable, in particular to improve the outcomes and effectiveness of the meetings of the Parties to the Cartagena and Nagoya protocols.

Increased Future Cooperation In 2016, Decision XIII/26, building on the inputs provided by a workshop held in Geneva in February 2016, considered options to enhance synergies among the biodiversity-related conventions and invited the Parties to further strengthen cooperation and coordination at the national (Annex I) and international (Annex II) levels. In 2018 the CBD CoP adopted Decision 14/30 urging Parties and other governments, organizations, conventions and stakeholders to consider possible new areas and approaches to advance the implementation of biodiversity commitments through enhanced cooperation as part of the post-2020 global biodiversity framework. In the section relating to cooperation with other conventions, the decision:

> Encourages consideration of actions for enhanced synergies among biodiversity-related conventions, the Rio Conventions, and other conventions that also address issues related to the three objectives of the Convention, in the development of the post-2020 global biodiversity framework, in particular as they are essential for the implementation of the 2030 Agenda for Sustainable Development and the Sustainable Development Goals. (para. 4)

Further integration of the biodiversity-related conventions is thereby recognized as a vital contribution to the post-2020 global biodiversity regime. However, the actions promoted so far by the governing bodies are essentially confined to improving administrative cooperation among the conventions' respective secretariats and bodies, whereas the important issue is to increase committed action by the States Parties. As has been rightly summarized,

> [t]he scale of cross-convention coordination [in the international biodiversity regime] exceeds cooperation in other areas of international environmental law, but it still pales in comparison to the task the conventions face; independent and uncoordinated action is still the norm, despite best efforts (Carlarne 2008, p. 469).

3.3.2 Conclusion of New Conventions

Gaps in the coverage of the international biodiversity regime can eventually be filled by the conclusion of new conventions to address non-regulated matters. This is the case with the ongoing process to conclude an international agreement concerning the regulation and management of marine biological diversity in areas beyond national jurisdiction.

Marine Biodiversity in Areas Beyond National Jurisdiction The issue of marine biodiversity witnessed a dramatic about-face in the late 1970s with the discovery that certain microorganisms and bacteria with unique genetic characteristics live near hydrothermal vents in the seabed in areas beyond national jurisdiction. The scientific and industrial sectors took note of the commercial potential of these seabed life forms, fuelling interest from many scientific laboratories, large multinationals and states that were drawn to the promising possibilities afforded by the discovery. The advancement of technology has enabled access to new genetic resources with extraordinary scientific and economic potential that is today open only to certain industrialized States.

Biological resources located in areas beyond national jurisdiction lack a specific global regime. The currently applicable general provisions, which are contained mainly in the LOS Convention and the CBD, offer a fragmentary and insufficient regulatory framework that can in no way respond to the present needs. In terms of the international law of the sea, the regime applicable to the exploration and exploitation of the genetic resources of the seabed is especially problematic. In fact, these resources are located in the contact zone between the deep waters of the high seas, where the principle of free access and exploitation governs (UNCLOS, article 87), and the international seabed area, whose legal status is defined as a "common heritage of mankind" (UNCLOS, article 136).[7] The jurisdictional scope of the Convention on Biological Diversity, as set out in article 3, is ambiguous: with respect to the "components" of biological diversity, the CBD applies only to areas within the limits of national jurisdiction. However, with respect to "processes and activities" carried out under the jurisdiction or control of each Contracting Party, the CBD applies both within and beyond areas subject to national jurisdiction. Article 5, which concerns the Contracting Parties' obligation to cooperate in matters of mutual interest for the conservation and sustainable use of biological diversity "also applies in respect of areas beyond national jurisdiction" (Vázquez Gómez 2019, p. 37).

Suitability of an International Legally Binding Instrument In the framework of various processes established by the CBD and the United Nations General Assembly, intensive work has been carried out on the suitability of developing an international instrument for the conservation and sustainable use of marine biodiversity beyond areas of national jurisdiction.

In 2004, the General Assembly agreed on the recommendations of the United Nations Informal Consultative Process on Ocean Affairs and the Law of the Sea (UNICPOLOS) to establish an *ad hoc* open-ended informal working group (the BBNJ Working Group) to study issues related to the conservation and sustainable use of marine biological diversity beyond areas of national jurisdiction. Between

[7]In article 136, the legal status of "common heritage of mankind" applies to the seabed area "and" its mineral resources. The issue of the legal status of the biological resources of the seabed area is not addressed in Part XI of UNCLOS.

2006 and 2010, however, little progress was made because of the group members' vast disparity of views on the most important issues. At its 2011 meeting, however, the BBJN Working Group reached a breakthrough by agreeing that the issues of protecting biodiversity should be linked "together and as a whole" with issues related to access and benefit sharing of marine genetic resources (Freestone 2019). On this understanding, the group recommended that "a process be initiated by the UNGA with a view to . . . the possible development of a multilateral agreement under the United Nations Convention on the Law of the Sea". The suggested process gained support in 2012 when the Rio+20 Summit adopted the document "The Future We Want" and the heads of state and government committed to address the issue "on an urgent basis" (para. 162). Thus, in its final recommendations in 2015, the BBJN Working Group stressed the need for a comprehensive global regime to better address the conservation and sustainable use of marine biological diversity in areas beyond national jurisdiction, and it recommended that the General Assembly "decide to develop an international legally binding instrument under UNCLOS" (para. 1.e).

On 19 June 2015, the General Assembly, by Resolution A/RES/69/292, endorsed the recommendation of the BBJN Working Group and decided to establish a preparatory committee to prepare the draft text of an international legally binding instrument for discussion by an international conference to be convened by the General Assembly before the end of its 72nd session in 2017. Once the preparatory committee completed its work, the General Assembly adopted Resolution 72/249 of 14 December 2017, whereby it decided:

> [. . .] to convene an intergovernmental conference, under the auspices of the United Nations, to consider the recommendations of the Preparatory Committee on the elements and to elaborate the text of an international legally binding instrument under the United Nations Convention on the Law of the Sea on the conservation and sustainable use of marine biological diversity of areas beyond national jurisdiction, with a view to developing the instrument as soon as possible.[8]

Ongoing Work of the Intergovernmental Conference The intergovernmental conference held its first organizational meeting on 16–18 April 2018, followed by three substantive sessions in 2019. The main topics of the ongoing negotiations [as agreed by the BBJN in 2011] are: marine genetic resources, including questions of benefit sharing; area-based management tools, including marine protected areas; the environmental impact assessment; and capacity-building and the transfer of marine technology. A new item agreed by the preparatory committee on general cross-cutting issues was also added to the list.

[8]The resolution also decided that the conference shall exhaust every effort in good faith to reach agreement on substantive matters by consensus and, if consensus cannot be reached, the decisions of the conference on substantive matters shall be taken by a two-thirds majority of the representatives present and voting (para. 17, 19). In practice, however, the spirit of the conference is to work towards consensus on a package deal for the future agreement.

Initially, the conference's discussions on the regime to govern the genetic resources of the seabed were polarized around two competing approaches, specifically between States promoting the idea of freedom of access and States advocating for exploitation in line with the principle that these resources are a common heritage of mankind (Scovazzi 2010). Gradually, this "ideological" debate gave way to a more pragmatic, comprehensive approach revolving around what developed States should provide to developing countries in terms of capacity-building, technology transfer, and access and benefit sharing (Tiller et al. 2019). After three sessions of negotiating, however, the delegations are still at loggerheads on crucial issues such as access to genetic resources and processes (*in situ*, *ex situ* or *in silico*), the sharing of non-monetary and monetary benefits, the spectrum of options for area-based management tools, capacity-building and the transfer of marine technology, and coherence between global and regional approaches. Another issue in dispute concerns the options for institutional arrangements, with the delegations taking positions that range from a preference for strong global institutions with decision-making powers, to a call for reliance on regional instruments, to a "hybrid" approach based on coordination and collaboration between new global institutions and existing regional and sectoral instruments (Oude Elferink 2019, p. 241).

A vast majority of delegations are now fully committed to the process and have submitted a large number of textual proposals. The draft text of an "effective, practicable and future proof" agreement is slowly developing as a valuable tool to address substantive matters and provide a solid basis for negotiations. However, as stated by the president of the conference at the close of the third session, much work is still required to streamline the text of the future agreement and find creative ways to spark solutions that may gain consensus among the negotiating delegations (UNGA 2019, p. 3).

The fourth and last session of the intergovernmental conference, which was set for the first half of 2020, has been postponed because of the Covid-19 emergency

3.3.3 *Operationalization of Principles*

A more substantial way to fill the current gaps in the biological diversity regime is to further operationalize the principles for the conservation of biodiversity which focus on the premise that it constitutes a collective interest of all States. Actually, the prevalence of the common interest is a common denominator among biodiversity conventions (Kim and Bosselmann 2013, pp. 290–291).

General Interest The recognition of the general interest of all States in the conservation of biological diversity has a long tradition in international law and jurisprudence.

The concept is implied in principle 2 of the 1972 Stockholm Declaration, which affirms that

> "the natural resources of the Earth, including flora and fauna and especially representative samples of natural ecosystems, must be safeguarded for the benefit of present and future generations."

In the same vein, the 1982 World Charter for Nature recognizes "the supreme importance" of protecting natural systems "in the interest of present and future generations". The concept of common interest also informs the International Whaling Convention (IWC 1949), the Convention on Wetlands (Ramsar 1971), the Convention on International Trade in Endangered Species of Wild Fauna and Flora (CITES 1973), the Bonn Convention on migratory species (MSC 1979) and a number of regional conventions. The preamble of the 1982 UNCLOS names the conservation of the living resources of the seas and oceans among the goals whose achievement "takes into account the interests and needs of mankind as a whole". The 2030 Agenda for Sustainable Development lists biodiversity loss as one of the greatest challenges undermining the ability of "all countries" to achieve sustainable development (para. 14).

Reference to the global interest in the protection of natural resources and biological diversity also appears in various international arbitral and judicial decisions. In the Bering Sea Arbitration of 1893, the United States claimed a right to restrain practices of pelagic sealing, alleging that there was a "common interest of all nations" in preventing the extermination of seals, which was against "the interest and welfare of mankind". The arbitral award, while discarding any right of protection or property of the US beyond its territorial sea, assumed the existence of a common interest of all nations in the protection and preservation of fur seals. The tribunal established a number of protective regulations that applied to both Parties while noting their agreement to cooperate in securing the adherence of other powers to such regulations.

The International Court of Justice (ICJ) has recognized the protection of the environment as an "essential interest" of all States in the Gabčikovo-Nagymaros Judgment (ICJ 1997, para. 53) and in the Pulp Mills Judgment, where the ICJ specifically referred to the "common interests dimension" of the protection of biodiversity (ICJ 2010, para. 184). The existence of a collective interest in the conservation of wildlife and biodiversity is implied in various judicial proceedings relating to the conservation and management of marine living resources brought before the ICJ (the ICJ 1998, Fisheries Jurisdiction Case -Spain v. Canada) and before the ITLOS (the 1999 Southern Bluefin Tuna Cases -Australia, New Zealand v. Japan- and the 2009 Swordfish Case-Chile/European Union).

In the case of whaling in the Antarctic brought before the ICJ, New Zealand contended that the system of collective regulation for the conservation and management of whale stocks established by the IWC requires that "the interests of all the parties in the conservation and management of whale stocks" be taken into account and balanced (ICJ 2014, para. 29). In its judgment on the dispute, the ICJ recalls that the first preambular paragraph of the International Convention for the Regulation of Whaling recognizes "the interest of the nations of the world in safeguarding for future generations the great natural resources represented by the whale stocks" (ICJ

2014, para. 56). In the same vein, the arbitral award of 12 July 2016 in the South China Sea case repeatedly refers to the interests of all States in the conservation of the living resources of the seas and marine biodiversity (CPA 2016, para. 956–959).

In these judicial and arbitral decisions, while the exercise of extraterritorial jurisdiction has been addressed in a rather restrictive way, the existence of a collective interest in the conservation of wildlife and biodiversity has been generally recognized.

Common Concern of Mankind The concept of a common concern of mankind, which was initially applied to certain areas of collective interest such as the Antarctic and outer space, has expanded to cover other global environmental values including biological diversity (Brunnée 2007, p. 564). Today, the concept inspires a number of environmental treaties and the term itself appears in the UN Convention on Climate Change (preamble, first paragraph).

Recognition of the collective interest in biodiversity conservation as a "common concern of humanity" is set out in article 3 of the 1995 IUCN Draft International Covenant on Environment and Development. The commentary on the article explains that the erosion of biological diversity adversely affects all humanity, and it calls for global solutions as a matter of "common concern". Likewise, the 2002 ILA New Delhi Declaration of Principles of International Law relating to sustainable development affirms that the protection, preservation and enhancement of the natural environment, particularly biological diversity and the fauna and flora of the Earth, "are the common concern of humankind" (article 1.3). In the preamble to the Convention on Biological Diversity, the parties emphatically proclaim that "the conservation of biological diversity is a common concern of humankind".[9] The principle of common concern of mankind is also stated in the preamble to the International Treaty on Plant Genetic Resources for Food and Agriculture (2001). The UNESCO Convention for the Protection of World Cultural and Natural Heritage uses a similar expression: "world heritage of mankind as a whole".

The operationalization of the concept of common concern to reflect the collective nature of States' interest in biodiversity protection might help to develop improved regulation and institutional governance. However, the implementation of the various consequences arising from the recognition of biodiversity as a common concern of mankind needs to undergo a difficult conceptual expansion, since it challenges "the foundational structures of international law" (Brunnée 2007, p. 553).

Equitable and Sustainable Use The principles of equitable and sustainable use of shared natural resources can provide additional support to the collective dimension of biodiversity conservation. With respect to activities relating to the exploitation of

---

[9]The reason for recognizing the principle of common concern of mankind in the preamble and not in the substantive provisions of the CBD can be related to the fact that the areas that still have an abundance of biodiversity are found mainly in the South, thus leading to pressure on developing States to conserve their environment in the interest not only of their own populations, but in the interest of the Earth's population as a whole (Hey 2016).

marine biological resources, especially fisheries, the principle was initially construed in a mostly utilitarian way requiring exploitation to be conducted at levels that were sustainable or optimal (Badenes Casino 1997).

However, international biodiversity law has progressively moved towards more ambitious goals. One of the main objectives of the CBD is the "fair and equitable sharing of the benefits arising out of the utilization of genetic resources" (article 1). The 2010 Nagoya Protocol on Genetic Resources establishes a regime for this purpose, but without qualifying what "fair and equitable share" entails other than that such sharing shall be "upon mutually agreed terms." In line with the tenets of the ICJ in the Gabčikovo-Nagymaros and Pulp Mills cases, it is "in relation to the allocation of shared natural resources that equity is likely to play an important role in coming years" (Sands and Peel 2012, p. 214).

Today, the concept of sustainable use of natural resources is undergoing an important evolution. The 2030 Agenda for Sustainable Development sets specific goals for the protection, restoration and promotion of sustainable use of marine resources and terrestrial ecosystems and for halting biodiversity loss (Goal 15), while also establishing other goals and targets in biodiversity-related fields (CBD et al. 2016). In so doing, the agenda stresses the need to "promote fair and equitable sharing of the benefits arising from the utilization of genetic resources" (15,6) and "mobilize significant resources from all sources to conserve and sustainably use biodiversity and ecosystems" (15,b).

***Erga Omnes* Obligation** In light of current developments in international environmental law, it could be argued that protection of the world's biological diversity is becoming an obligation owed to the international community as a whole (*erga omnes*) (Birnie et al. 2009, p. 131).

Under this assumption, the 2001 ILC Draft Articles on the responsibility of states for international wrongful acts may provide support for claims concerning the breach by a State of an obligation concerning the conservation of biodiversity owed to a group of States (*erga omnes partes*) or to the international community as a whole (*erga omnes*) (article 42). In such cases, an injured State or non-injured State may, under some conditions, invoke the responsibility of another State for breach of its collective obligations relating to the conservation of biodiversity (article 48). As stated in the ILC commentary to article 48, in such cases the principal purpose of the State invoking the responsibility of another State "will be to foster a common interest, over and above any interests of the States concerned individually" (ILC 2001, p. 126).

Most conventions on biodiversity set up collective obligations whose breach entitles other Parties, whether injured or not injured, to invoke the responsibility of the offending State. Moreover, in cases where a jurisdictional link exists, an injured or non-injured State could bring a claim for breach of an obligation *erga omnes partes* relating to biodiversity before a competent arbitral or judicial tribunal. Such jurisdictional links are actually provided for in many biodiversity conventions, including the Convention on Biological Diversity and its Protocols (article 27), the

Bonn Convention on Migratory Species (article XIII), CITES (article XVIII) and the UNCLOS provisions relating to the settlement of disputes by jurisdictional means (Part XV, Section 2). In sum, if the requirements laid down in those jurisdictional clauses are met, a State party to such biodiversity conventions may bring proceedings before a competent arbitral or judicial court against another party that has breached its conventional obligations, even if the applicant State is not directly injured by the violation. Prominent examples confirming the locus standi of States parties to biodiversity-related conventions for breaches of obligations *erga omnes partes* are the ICJ 2014 Judgement on the Whaling in the Antarctic case (Australia v. Japan, New Zealand intervening) and the 2016 CPA arbitral award in the South China Sea case (the Philippines v China).

The same situation may arise when an *erga omnes* obligation owed to the international community as a whole is breached, although there is not yet any confirmative jurisdictional practice. This interpretation has been endorsed by the International Law Institute's resolution on "*erga omnes* obligations and rights under international law" adopted in 2005 (Institut de droit international, Krakow session 2005, Art. 3).

3.4 Conclusion

The persistence of gaps in the international biodiversity regime on the eve of its post-2020 phase is an indisputable fact. What is worse, the unremitting decline of biodiversity that challenges the future of the Earth system is plain to see. The IPBES's 2019 Global Assessment Report on Biodiversity and Ecosystem Services predicts that over a million species, including 33% of reef-forming corals and marine mammals, could disappear entirely in our lifetimes (IPBES 2019, point 6).

In order to stop this from happening, there is a need to act with a sense of urgency and determination to find practical ways to bridge existing gaps. Thus far, the actions that have been taken focus on policies that seek to enhance scientific and technical cooperation and administrative coordination among the biodiversity-related conventions. Certainly, such efforts to improve the implementation of biodiversity conventions in a mutually supportive manner by developing linkages among them offer a promising way forward. However, if the loss of biodiversity is to be halted, States will also need to undertake more committed action to enhance cooperation, ensure coherence and provide for an efficient international legal regime to preserve the Earth's biodiversity.

In recent years, ambitious doctrinal calls have been made to merge existing biodiversity conventions by developing them into distinct protocols under the CBD (Jóhannsdóttir et al. 2010, pp. 147–148). Such calls, however, have fallen on deaf ears and their implementation seems unrealistic at present. The piecemeal amalgamation of disparate conventions appears to be a structural legal feature that will remain unchanged for some time to come. In fields of particular economic interest for States, new convention instruments could eventually be concluded to fill material gaps, such as the impending agreement on the conservation and sustainable use of marine biodiversity in areas beyond national jurisdiction. With the set-back of the process towards a Global Pact for the Environment in the form of a treaty in 2019, however, no other structural changes appear to have a realistic prospect of success.

Under this scenario, in order to overcome the challenges of the post-2020 framework the respective CoPs to the biodiversity conventions have a real responsibility to adopt ambitious decisions that commit the Parties to take effective action for the conservation and sustainable development of the Earth's biodiversity.

References

Badenes Casino, M. (1997). *La crisis de la libertad de pesca en alta mar*. Madrid: McGraw Hill.

Birnie, P., Boyle, A., & Redgwell, C. (2009). *International law & the environment*. Oxford: Oxford University Press.

Brunnée, J. (2007). Common areas, common heritage, and common concern. In D. Bodansky, J. Brunnée, & E. Hey (Eds.), *The Oxford handbook of international environmental law* (pp. 550–573). Oxford: Oxford University Press.

Cardesa-Salzmann, A. (2017). Monitoring and compliance mechanism. In J. Razzaque & E. Morgera (Eds.), *Biodiversity and nature protection law (Chapter 33), Elgar Encyclopedia of Environmental Law Series*. Edward Elgar Publishing.

Carlarne, C. P. (2008). Good climate governance: Only a fragmented system of international law away? *Law & Policy, 30*(4), 450–480.

CBD, FAO, the World Bank, UN Environment, and UNDP. (2016). *Biodiversity and the 2030 Agenda for Sustainable Development*. Policy Brief; Technical Note. Retrieved September 29, 2001, from https://www.undp.org/content/undp/en/home/librarypage/environment-energy/ecosystems_and_biodiversity/biodiversity-and-the-2030-agenda-for-sustainable-development%2D%2D-p.html

Chambers, W. B. (2008). *Interlinkages and the effectiveness of multilateral environmental agreements*. Tokyo–New York–Paris: United Nations University Press.

Dunne, R. P., Polunin, N. V. C., Sand, P. H., & Johnson, M. L. (2014). The creation of the Chagos marine protected area: A fisheries perspective. In M. L. Johnson & J. Sandell (Eds.), *Advances in marine biology* (Vol. 69, pp. 79–127). Oxford: Academic Press.

Edeson, W. R. (2016). Practice of compliance control within the framework of the protection of marine living resources. In U. Beyerlin, P.-T. Stoll, & R. Wolfrum (Eds.), *Ensuring compliance with multilateral environmental agreements* (pp. 151–173). Leiden/Boston: Martinus Nijhoff Publishers.

Freestone, D. (Ed.). (2019). *Conserving biodiversity in areas beyond national jurisdiction*. Leiden/Boston: Martinus Nijhoff Publishers.

French, D., & Kotzé, L. J. (2019). Towards a global pact for the environment: International environmental law's factual, technical and (unmentionable) normative gaps. *Review of European, Comparative and International Environmental Law, 28*(1), 25–32.

Hey, H. (2016). *Advanced introduction to international environmental law*. London: Edward Elgar Publishing.

ICJ. Gabčikovo-Nagymaros Project (Hungary/Slovakia), Judgment, ICJ. Reports 1997, p. 7.

ICJ. Fisheries Jurisdiction (Spain v. Canada), Jurisdiction of the Court, Judgment, ICJ Reports 1998, p. 432.

ICJ. Pulp Mills on the River Uruguay (Argentina v. Uruguay), Judgment, ICJ Reports 2010, p. 14.

ICJ. Whaling in the Antarctic (Australia v. Japan: New Zealand intervening), Judgment, ICJ Reports 2014, p. 226.

ILC. International Law Commission, *Draft Articles on Responsibility of States for Internationally Wrongful Acts*, November 2001, Supplement No. 10 (A/56/10), Chap. IV.E.1. Also available at: Draft articles on Responsibility of States for Internationally Wrongful Acts, with commentaries, 2001. *Yearbook of the International Law Commission*, 2001, Vol. II, Part Two.

International Law Institute, Krakow session – 2005. Fifth Commission. Obligations and rights erga omnes in international law, Rapporteur: Giorgio Gaja. Adopted on 27 August 2005.

ITLOS. Southern Bluefin Tuna (New Zealand v. Japan; Australia v. Japan), Provisional Measures, Order of 27 August 1999, ITLOS Reports 1999.

ITLOS. Conservation and Sustainable Exploitation of Swordfish Stocks (Chile/European Union), Order of 16 December 2009, ITLOS Reports 2008–2010.

Jeffrey, M., Freestone, J., & Bubna-Litic, K. (2008). *Biodiversity conservation, law and livelihoods: Bridging the North–South Divide*. Cambridge: IUCN Academy of Environmental Law Research Studies.

Jóhannsdóttir, A., Cresswell, I., & Bridgewater, P. (2010). The current framework for international governance of biodiversity: Is it doing more harm than good? *Review of European Community and International Environmental Law, 19*(2), 139–149.

Juste Ruiz, J. (2016). Création par le Royaume-Uni d'une aire marine protégée autour de l'archipel des Chagos. Différend entre le Royaume-Uni et l'Ile Maurice. Procédure arbitrale fondée sur la Convention des Nations Unies sur le droit de la mer. *Revue Juridique de l'Environnement, 1*, 141–152.

Juste Ruiz, J. (2018). Ocean options for climate change mitigation: Disposal of greenhouse gases at the sea under the 1996 London Protocol. In Rosa María Giles Carnero (coord.) *Desafíos de la acción jurídica internacional y europea frente al cambio climático* (pp. 43–56). Barcelona: Atelier.

Juste Ruiz, J. (2020). The process towards a global pact for the environment: From legal ambition to political dilution. *Review of European Community and International Environmental Law, 2020*, 1–12. https://doi.org/10.1111/reel.12331

Kim, R. E., & Bosselmann, K. (2013). International environmental law in the Anthropocene: Towards purposive system of multilateral environmental agreements. *Transnational Environmental Law, 2*(2), 285–310.

Kiss, A. (2006). Reporting obligations and assessment of reports. In U. Beyerlin, P.-T. Stoll, & R. Wolfrum (Eds.), *Ensuring compliance with multilateral environmental agreements* (pp. 229–245). Leiden/Boston: Martinus Nijhoff Publishers.

Oude Elferink, A. G. (2019). Exploring the future of the institutional landscape of the oceans beyond national jurisdiction. *Review of European Community and International Environmental Law, 28*, 236–243.

PCA Case N° 2013-19, In the matter of The South China Sea Arbitration before an Arbitral Tribunal Constituted under Annex VII to The1982 United Nations Convention on The Law of the Sea, between the Republic of the Philippines and the People's Republic of China, Award of 12 July 2016.

Prieur, M. (2018). *Les indicateurs juridiques. Outils d'évaluation de l'effectivité du droit de l'environnement*. Québec: Institut de la Francophonie pour le Développement Durable (IFDD).

Rockström, J., Steffen, W., Noone, K., Persson, Å., Chapin III, F. S., Lambin, E., et al. (2009). A safe operating space for humanity. *Nature, 461*(7263), 472–475.

Sands, P., & Peel, J. (2012). *Principles of international environmental law*. Cambridge: Cambridge University Press.

Scovazzi, T. (2004). Marine protected areas on the high seas: Some legal and policy consideration. *The International Journal of Marine and Coastal Law, 19*(1), 1–17.

Scovazzi, T. (2010). Is the UN Convention on the law of the sea the legal framework for all activities in the sea? The case for bioprospecting. In D. Vidas (Ed.), *Law, technology and science for oceans in globalisation* (pp. 309–317). Leiden–Boston: Martinus Nijhoff Publishrs.

Scovazzi, T. (2014). Marine protected areas in waters beyond National Jurisdiction. In M. Chantal Ribeiro (Coord.), *30 Years after the signature of the United Nations* Convention on the law of the sea (pp. 209–238). Coimbra: Coimbra Editora.

Tiller, R., De Santo, E., Mendenhall, E., & Nyman, E. (2019). The once and future treaty: Towards a new regime for biodiversity in areas beyond national jurisdiction. *Marine Policy, 99*, 239–242.

UNGA. (2019, September 13). Statement by the President of the conference at the closing of the third session, A/CONF.232/2019/10*.

UNGA. Transforming our world: the 2030 Agenda for Sustainable Development, UN Doc. A/RES/70/1.

UNGA. Letter dated 13 February 2015 from the Co-Chairs of the Ad Hoc Open-ended Informal Working Group to the President of the General Assembly. Annex, Outcome of the Ad Hoc Open-ended Informal Working Group to study issues relating to the conservation and sustainable use of marine biological diversity beyond areas of national jurisdiction and Co-Chairs' summary of discussions, UN Doc. A/69/780.

United Nations. Reports of International Arbitral Awards. Award between the United States and the United Kingdom relating to the rights of jurisdiction of United States in the Bering's sea [sic] and the preservation of fur seals 15 August 1893. Volume XXVIII, pp. 263–276.

Vadrot, A. B. M. (2014). *The politics of knowledge and global biodiversity*. London–New York: Routledge.

Vázquez Gómez, E. (2019). La protección de la diversidad biológica marina más allá de la jurisdicción nacional. hacia un nuevo acuerdo de aplicación de la Convención de Naciones Unidas sobre el derecho del mar. *Revista Electrónica de Estudios Internacionales*, 1–29.

Chapter 4
Opportunities and Challenges for Synergies Across Biodiversity-Related Conventions in the Context of Human Health and Zoonotic Diseases: The Role of Scientific Advisory Bodies

Royal C. Gardner

Contents

4.1 Introduction 58
4.2 Overview of Scientific Advisory Bodies 59
4.2.1 Ramsar Convention on Wetlands 59
4.2.2 Convention on International Trade in Endangered Species (CITES) 60
4.2.3 Convention on the Conservation of Migratory Species of Wild Animals (CMS) 62
4.2.4 Agreement on the Conservation of African-Eurasian Migratory Waterbirds (AEWA) 63
4.2.5 Convention on Biological Diversity (CBD) 64
4.2.6 Collaboration Among the Scientific Advisory Bodies 65
4.3 IPBES (and Its Relationship with Biodiversity-Related Conventions) 66
4.4 Scientific Advisory Bodies' Responses to Outbreaks of Specific Zoonotic Diseases 66
4.4.1 Highly Pathogenic Avian Influenza (HPAI) 67
4.4.2 Ebola Virus Disease (Ebola) 68
4.4.3 COVID-19 69
4.5 Conclusion 71
References 72

Abstract This chapter examines the work of the scientific advisory bodies of five biodiversity-related conventions: the Ramsar Convention, Convention on International Trade in Endangered Species, Convention on Migratory Species, Agreement on the Conservation of African-Eurasian Migratory Waterbirds, and Convention on Biological Diversity. It will discuss how the scientific advisory bodies have collaborated in the past, through ad hoc or more formalized approaches, and how they may work more closely together in the future through the emergence of the

---

I would like to acknowledge the outstanding work done by my research assistant Lauren Beames.

R. C. Gardner (✉)
Stetson University, College of Law, Gulfport, FL, USA
e-mail: GARDNER@law.stetson.edu

M. Campins Eritja, T. Fajardo del Castillo (eds.), *Biological Diversity and International Law*, https://doi.org/10.1007/978-3-030-72961-5_4

Intergovernmental Science-Policy Platform on Biodiversity and Ecosystem Services (IPBES), especially in the context of human health and infectious diseases. In doing so, this chapter considers how the scientific advisory bodies have responded to outbreaks of highly pathogenic avian influenza, Ebola virus disease, and COVID-19. While institutional constraints exist, IPBES provides an opportunity for the scientific advisory bodies—and thus the biodiversity-related conventions themselves—to more effectively emphasize the need for biodiversity conservation as preventive measures for human health and well-being.

4.1 Introduction

Biodiversity and human health are inherently linked in innumerable ways. One particular connection involves the spread of zoonotic and vector-borne diseases. The destruction and degradation of the natural environment, as well as human-wildlife interactions, increase the risk of such diseases (WHO and Secretariat of the CBD 2020). The Manhattan Principles on "One World, One Health" recognized the relationship between human, domestic animal, and wildlife health, noting that effective responses to disease threats should "fully integrate biodiversity conservation perspectives and human needs" (Cook et al. 2004).

Up to 75% of new or emerging infectious diseases are zoonotic in origin (Jones et al. 2008). The current COVID-19 global pandemic that spread in the wet markets of Wuhan provides the most recent and starkest example, as the coronavirus SARS-CoV-2, which causes COVID-19, has a zoonotic origin. Bat populations are suspected to be the source, although this has not been confirmed (WHO 2020).

Sustainable Development Goal (SDG) 3, which calls for ensuring healthy lives and promoting well-being, contains combating communicable diseases in Target 3.3 (United Nations (UN) 2015). Often, however, biodiversity conservation is overlooked as a component of human health and well-being (Convention on Biological Diversity 2017). The SDGs contemplate that action needs to be taken at all scales, from the local to the international. To what extent do biodiversity-related conventions—and in particular their scientific advisory bodies—contribute, individually and collectively, to efforts to reduce the risks of outbreaks of infectious diseases?

This chapter will examine the structure, roles, and responsibilities of the scientific advisory bodies of five biodiversity-related conventions: the Ramsar Convention, Convention on International Trade in Endangered Species, Convention on Migratory Species, Agreement on the Conservation of African-Eurasian Migratory Waterbirds, and Convention on Biological Diversity. It will focus on how these scientific bodies advise convention bodies, convention parties, and broader stakeholders. It will also note how the scientific advisory bodies have collaborated in the past, through ad hoc or more formalized approaches, and how they may work more closely together in the future through the emergence of the Intergovernmental Science-Policy Platform on

Biodiversity and Ecosystem Services (IPBES), especially in the context of human health and infectious diseases. In doing so, this chapter will consider how the scientific advisory bodies have responded to outbreaks of highly pathogenic avian influenza, Ebola virus disease, and COVID-19. While institutional constraints exist, there is an opportunity for the scientific advisory bodies—and thus the biodiversity-related conventions themselves—to more effectively emphasize the need for biodiversity conservation as preventive measures for human health and well-being.

4.2 Overview of Scientific Advisory Bodies

Scientific advisory bodies are a common feature of biodiversity-related conventions. Traditionally, and to some extent still, they operate in silos independent of each other. Part of the reason, of course, is that that their mandates relate to the primary conservation objectives of their respective conventions and their principal audiences are their respective convention's constituencies. Yet their conservation objectives and constituencies overlap, especially as the conventions make the linkages between biodiversity conservation, zoonotic diseases, and human health. These overlaps suggest an opportunity for synergies and collaboration among the scientific advisory bodies. Accordingly, the scientific advisory bodies occasionally work and coordinate their actions with each other. At one time, the Chairs of the Scientific Advisory Bodies would meet to discuss work plans and possible areas of collaboration, but this mechanism has recently fallen into disuse. Without sustained institutional support, collaboration largely has depended on personal efforts and relationships.

4.2.1 Ramsar Convention on Wetlands

Concluded in 1971, the Ramsar Convention, which promotes conservation and wise use of wetlands, is one of the earliest multilateral biodiversity conventions (Gardner and Davidson 2011). Wetlands are defined very broadly to include marine/coastal, inland, and human-made aquatic ecosystems, ranging from swamps, bogs, marshes, and fens to coral reefs, rivers, lakes, reservoirs, and rice fields (Finlayson and Davidson 2018). An overarching principle is the wise use concept: parties are expected to maintain the ecological character of the wetlands on their territories. Part of a wetland's ecological character is its species (Gardner and Finlayson 2018). Although the Convention began with an emphasis on wetlands as waterfowl habitat (indeed, that was its impetus), its scope has broadened beyond wetland-dependent species to consider how wetlands contribute to all aspects of sustainable development – including areas such as poverty elimination and human health.

The Ramsar Convention's scientific advisory body, the Scientific and Technical Review Panel (STRP), was established by Resolution 5.5 (Ramsar Convention 1993). Under its current *modus operandi*, the STRP has 18 members (Ramsar

Convention 2015), with the aim of regional representativeness. STRP members serve in their personal capacities; they are expected to provide unbiased advice and do not represent a particular country.

The STRP is supplemented and supported by expertise from observer organizations and the Contracting Parties' STRP National Focal Points. Many observer organizations have a primary focus on wetlands. But observers may also include the chairs of the scientific advisory bodies of other biodiversity-related conventions, as well as non-wetland centred entities such as the Food and Agriculture Organization and World Health Organization (Ramsar Convention 2015).

Advice and guidance from the STRP may take several forms. The STRP has developed guidance that has been attached as annexes to resolutions adopted by the Conference of the Parties (CoPs), including in the area of wetlands and human health (Ramsar Convention 2012). The STRP also provides guidance through its publications such as technical reports and briefing notes, some of which have covered the interplay between healthy wetlands and healthy people (Horwitz et al. 2012; Cromie et al. 2012).

One challenge for the STRP is its workload. Throughout its history, requests of the STRP have outpaced its capacity (Ramsar Convention 1996). Currently, the STRP develops a draft work plan that identifies high priority tasks for the approval of the Standing Committee, the Convention's decision-making body during the intersessional periods between CoPs (Ramsar Convention 2015). As is traditional in Ramsar, the Standing Committee strives to reach decisions by consensus. Thus, a single country or group of countries can block (or water down) decisions. For example, during a Zika outbreak, the STRP recommended providing advice regarding mosquito disease vectors in wetlands for site managers, practitioners, and policy makers as a high priority task (Ramsar Convention 2016a). After several Contracting Parties, led by Brazil, questioned whether this issue was within the STRP's remit, the task was revised to focus generally on wetland management to reduce the risk of waterborne and aquatic-vector borne diseases (Ramsar Convention 2016b). The task was also dropped from the high priority list and consequently was not delivered.

4.2.2 Convention on International Trade in Endangered Species (CITES)

CITES, which was concluded in 1973 and entered into force in 1975, was developed to respond to the over-exploitation of animal and plant species through international trade (Favre 1989). CITES regulates, but does not prohibit, international trade in endangered species. The goal of CITES is legal, sustainable, and traceable trade, which is sought to be achieved primarily through a licensing regime. Species are listed on one of three appendices; the trade of Appendix I species is the most restricted.

CITES has two scientific advisory bodies, the Animals Committee and the Plants Committee, both of which were established in 1987 at the CoP6 (CITES 2020a). The Animals Committee, which is more relevant for this chapter, has ten regionally-based members. A zoological nomenclature expert is a non-voting, *ex officio* member. Similar to Ramsar STRP members, Animal Committee members are expected to "act as impartially as possible and endeavour to base his/her judgements and opinions upon an objective, scientific assessment of the available evidence" (CITES 2019a).

Parties and UN agencies are observers and may send representatives to committee meetings. Moreover, the Rules of Procedure specify that, with the permission of the Chair of the Animals Committee, observers can be "[a]ny body or agency technically qualified in protection, conservation or management of wild fauna and flora" (CITES 2018a). This encompasses non-governmental agencies and bodies, including private sector entities. Thus, in addition to animal welfare organizations, Animals Committee meetings are attended by groups representing the interests of the pet trade, trophy hunters, and food importers (CITES 2018b).

Consistent with instructions from the CoP, the Animals Committee provides scientific advice and guidance "on matters relevant" to the international trade of animal species listed in the Appendices (CITES 2019a). In particular, the Animals Committee may draft resolutions for consideration by the CoP; advise on nomenclature issues; periodically examine Appendix I and II to ensure that species are appropriately listed; and conduct Reviews of Significant Trade in specimens of Appendix II species, which may include recommendations on how to respond to unsustainable trade. Given CITES' focus on international trade, human health issues have not figured prominently in the Animals Committee's work (CITES 2012a).

Nevertheless, human health is considered and is relevant to some aspects of CITES' agenda. For example, the CoP adopted a resolution on wild meat (bushmeat) and established a Central Africa Bush Meat Working Group (now non-operational) (CITES 2019b). In light of the connection between wild meat and zoonotic diseases (Wolfe et al. 2005), it would appear that CITES could consider human health impacts. In this context, however, it seems that human health is only taken into account to the extent that the regulation of trade of wild meat affects livelihoods, not with regard to the risk that this trade may increase the prevalence of zoonotic diseases (Van Vliet et al. 2016).

In 2012, the Animals Committee did expressly consider the linkages between CITES and wildlife diseases, including zoonotic diseases (CITES 2012a). However, it concluded "that wildlife diseases were best addressed by other organizations, such as the World Organization for Animal Health (OIE)" (CITES 2012b).

In 2015, the CITES Secretariat entered into a cooperation agreement with the OIE (CITES and OIE 2015). The agreement primarily contemplates collaboration concerning animal health and welfare standards related to the international transport of animals included in CITES Appendices. The agreement also anticipates CITES' collaboration with the OIE Working Group on Wildlife Diseases, which has in its terms of reference a mandate to "[m]aintain global perspective" on, *inter alia*, zoonoses (OIE 2020). However, a recent Secretariat statement on COVID-19 (discussed further below) takes the position that issues related to zoonotic diseases are beyond CITES' mandate and competency (CITES 2020b).

4.2.3 Convention on the Conservation of Migratory Species of Wild Animals (CMS)

Concluded in 1979 and entering into force in 1983, CMS promotes the conservation of migratory species, their habitats, and their migratory routes (Bowman et al. 2010). Species threatened with extinction are listed on Appendix I based on the "best scientific evidence available," and the Convention imposes obligations on Range States regarding such species, including strictly limiting their taking (CMS 1979). Appendix II species are those with an "unfavourable conservation status" that could benefit from international cooperation. The Convention encourages Range States to enter into agreements and more informal memoranda of understanding (MOUs) to improve the status of these species. Thus, there is a "CMS Family" of agreements and MOUs pertaining to specific species. (One such instrument, the Agreement on the Conservation of African-Eurasian Migratory Waterbirds, is discussed below.)

Article VIII of the Convention provides for a Scientific Council to be established at the first CoP. It further states that "[a]ny Party may appoint a qualified expert as a member of the Scientific Council," and thus the size of the Scientific Council is a function of the number of Parties. The Scientific Council is expected to provide scientific advice regarding migratory species to CMS Parties and bodies, as well as those of other agreements within the CMS Family.

As the number of CMS Parties grew, it became apparent that the functioning of the Scientific Council needed to be improved. A review found that expertise on the Scientific Council was unbalanced, with an over-emphasis on avian experts (CMS 2014). Given the size of the Scientific Council, budget constraints limited its intersessional meetings and little work progressed between CoPs. When the Scientific Council did meet, it functioned like a mini-CoP. In plenary, Scientific Councillors sat behind the name plates of their countries. When recognizing them to speak, the Scientific Council Chair would refer to their country's name, thus suggesting that the intervention represented the views of the Party (and not necessarily the independent views of the individual Councillor).

To increase the effectiveness of the Scientific Council, CoP11 established the Sessional Committee of the Scientific Council, which was further refined by CoP12 (CMS 2017). The Sessional Committee is smaller, composed of nine members with expertise in taxonomic and thematic issues and fifteen regionally-based members. Representatives of the scientific advisory bodies of multilateral environmental agreements, including those within the CMS Family, are invited to attend Scientific Council and Sessional Committee meetings as observers.

In contrast to CITES, CMS considers that its mandate includes issues related to zoonotic diseases and human health. If a country's response to a zoonotic outbreak is to cull the wild animals believed to be the origin and to destroy their habitat, CMS obligations may be implicated (and violated). Accordingly, CMS is a co-convener with FAO of the Scientific Task Force on Wildlife and Ecosystem Health, which tracks outbreaks of, *inter alia*, highly pathogenic avian influenza and provides advice on appropriate response actions (CMS 2008).

4.2.4 Agreement on the Conservation of African-Eurasian Migratory Waterbirds (AEWA)

AEWA, one of the agreements under the CMS umbrella, was adopted in 1995 and entered into force in 1999 (Lewis 2016). Parties commit to maintaining the favourable conservation status of migratory waterbirds, or restoring them to that status, through an Action Plan that is annexed to and an integral part of the agreement. Significantly, the agreement expressly provides the precautionary principle as a fundamental principle (AEWA 2018a).

Article VII of AEWA established a Technical Committee as its scientific advisory body. The Technical Committee is comprised of nine regional experts (one each from AEWA's nine regions) and one expert each in the areas of rural economics, game management and environmental law. In addition, the Technical Committee has one representative each from IUCN, Wetlands International, and International Council for Game and Wildlife Conservation. While Parties may send observers to Technical Committee meetings, observer participation from organizations is more limited than in other biodiversity-related conventions. Article VII of the agreement specifies that the Technical Committee Chair may admit no more than four observers from international inter-governmental and non-governmental organizations.

The Technical Committee advises the Meeting of the Parties, as well individual Parties, on matters related to AEWA's species and flyway conservation. In particular, the Technical Committee makes recommendation on the Action Plan (AEWA 2018b). The Technical Committee's responsibilities also include advising on emerging issues, which can relate to zoonotic diseases. As was the case with the Ramsar STRP, the tasks requested of AEWA's Technical Committee grew over time with no corresponding increase in budget support (AEWA 2018c). Accordingly, the Meeting of the Parties now identifies top priority tasks (those deemed "essential" or "high") in the work plan on which the Technical Committee should focus its efforts.

Due to the connection of wild birds and highly pathogenic avian influenza, one high task, which has rolled over from previous work plans, relates to emerging diseases. The task calls for the Technical Committee to "[c]ontinue to participate in the CMS Scientific Council Working Group on wildlife diseases" (AEWA 2018c). As discussed below, AEWA's Technical Committee has played a significant role in developing guidance on responses to outbreaks of highly pathogenic avian influenza.

4.2.5 Convention on Biological Diversity (CBD)

Concluded at the 1992 United Nations Conference on Environment and Development in Rio de Janeiro and entering into force only a year later, CBD has the broadest scope of the biodiversity-related conventions (Secretariat of the CBD 2005). Its three primary objectives are to: conserve biodiversity; sustainably use the components of biodiversity; and fairly and equitably share the benefits from the use of genetic resources.

Similar to CMS and AEWA, CBD expressly refers to a scientific advisory in its convention text. Article 25 establishes the Subsidiary Body on Scientific, Technical and Technological Advice (SBSTTA), which provides advice to the CoP as well as subsidiary convention bodies (United Nations (UN) 1992). Article 25 further specifies that the multidisciplinary SBSTTA "shall be open to participation by all Parties" and consist of "government representatives competent in the relevant field of expertise." Thus, SBSTTA members represent parties and do not necessarily provide independent scientific advice. Moreover, because CBD has near universal membership (196 parties), the size of SBSTTA is huge in comparison to the other scientific advisory bodies. In addition, representatives of various observer organizations, including the scientific advisory bodies of other biodiversity-related conventions, may attend SBSTTA meetings.

SBSTTA generally reviews assessments and guidance produced by others, such as, for example, UNEP-WCMC, *ad hoc* technical expert groups, and the Secretariat. In the health context, for example, at the request of CoP13, the Secretariat prepared guidance on integrating biodiversity considerations into One Health approaches (CBD 2017). After the SBSTTA review, CoP14 endorsed the guidance (CBD 2018a). The guidance observed that "land use change (e.g., land conversion, oil extraction, mining, deforestation, wetland drainage, etc.) has been identified as the leading driver of infectious disease emergence from wildlife." It encourages the use of the One Health approach to

> minimiz[e] unnecessary disturbance to natural systems and so avoid or mitigate the potential emergence of new pathogens, and manage the risk of transmission of pathogens between humans, livestock and wildlife in order to reduce the risk and incidence of infectious diseases, including zoonotic and vector-borne diseases.

Protection of biodiversity for human health benefits is thus firmly embedded in CBD's agenda.

Given its composition, however, SBSTTA operates more like a political rather than a scientific body. As Urho (2009) noted, SBSTTA "has developed into what many refer to as a 'Mini-CoP' – a body with strong political features negotiating draft decisions in preparation for the subsequent meetings of the CoP." Urho's observation that a mere 7% of SBSTTA delegates are directly associated with academic and research institutions underscores the political nature of SBSTTA.

4.2.6 Collaboration Among the Scientific Advisory Bodies

Because of the biodiversity-related conventions' overlapping objectives, the potential positive synergies of collaboration, including with regard to science-policy interface, has long been recognized (UNEP 2015). Occasionally, scientific advisory bodies, such as the Ramsar STRP and AEWA's Technical Committee, have coordinated aspects of their work programmes. Other examples, albeit also on an *ad hoc* basis, are joint guidance documents and technical reports (de Groot et al. 2006; Gitay et al. 2011). Sometimes these collaborative efforts have been the result of overlapping appointments (i.e., one person serving on two bodies) or personal relationships.

Limited capacity in terms of financial resources and time is a primary challenge to greater collaboration among the scientific advisory bodies. In concept, the various meetings of the scientific advisory bodies invite representatives from their sister scientific advisory bodies to attend as observers. In practice, participation is sporadic. The budgets for the scientific advisory bodies generally do not support regular travel to the meetings of other scientific advisory bodies. Even if funds are available, time is a constraint; the availability of representatives of a scientific advisory body to attend the meeting of another is often limited due to other professional commitments, especially if service on a body is volunteer-based.

One systematic attempt at collaboration was the Chairs of the Scientific Advisory Bodies (CSAB). Mandated by the Biodiversity Liaison Group (comprised of representatives from the secretariats of the biodiversity-related conventions), CSAB first met in 2007 with the goal of identifying "areas of cooperation and collaboration on the scientific issues of the various convention processes and their translation into policy" (UNEP-WCMC 2012). CSAB considered possible joint projects and programmes related to current priorities in their work programmes, such as

> "guidance on ecosystem restoration; specific climate change-related questions; hunting and harvesting; tourism and ecotourism; urbanization; and invasive species" (UNEP and WCMC 2012).

Regrettably, however, CSAB has not delivered on its objectives. One reason is that it lacks formal administrative support from the secretariats (Gardner and Grobicki 2016). For example, although the CITES Secretariat offered to "provide to CSAB Members log-in data for an intersessional discussion forum, hosted on its website" (CBD 2013), such a forum was not established. Similarly, the CBD Secretariat offered to host a CSAB meeting in conjunction with SBSTTA-18 in 2014, but the meeting did not take place. Accordingly, CSAB has not convened since its sixth meeting in Formia, Italy, in 2013 and is now in what might be diplomatically called a quiet phase (Gardner and Grobicki 2016).

Although CSAB is not currently functional, the biodiversity-related conventions participate in the Intergovernmental Science-Policy Platform on Biodiversity and Ecosystem Services (IPBES). This offers another avenue for collaboration and cooperation.

4.3 IPBES (and Its Relationship with Biodiversity-Related Conventions)

Established in 2012, IPBES is an independent intergovernmental body designed to strengthen the policy-science interface (Díaz et al. 2015). As its name suggests, its focus is on both the conservation and sustainable use of biodiversity and on nature's contributions to people and human well-being. IPBES is best known for its assessments, especially its global assessment that warned that a million species are at risk of extinction (Intergovernmental Science–Policy Platform on Biodiversity and Ecosystem Services (IPBES) 2019).

The Plenary, which essentially operates as a CoP, meets annually and approves the IPBES work programme. A Multidisciplinary Expert Panel (MEP), the members of which are elected by the Plenary on a regional basis, provides advice on the work programme and selects the experts who produce IPBES deliverables. Eighty per cent of the experts must come from the nominations of governments; the remaining 20% may come from the nominations of scientific organizations, which include the biodiversity-related scientific advisory bodies (Hotes and Opgenoorth 2014).

The biodiversity-related conventions are IPBES stakeholders and participate significantly in the IPBES process. Ramsar, CITES, CMS, and CBD each have a memorandum of cooperation with IPBES (IPBES 2020a). They submit proposals for the work programme, nominate experts to serve as authors, develop data that feed into IPBES assessments, and review draft IPBES products. The biodiversity-related conventions are also viewed as one of the end-users of IPBES products.

Furthermore, the chairs of the scientific advisory bodies are observers to the MEP (IPBES 2020b). Chairs (or their representatives) of the scientific bodies of Ramsar, CITES, CMS, and CBD have all attended MEP meetings and IPBES plenaries, which suggests an opportunity for collaboration at least on the margins of these events. However, the time and resource constraints on travel for CSAB meetings apply in this context as well, and thus far not all the chairs have been able to attend the same IPBES meeting. Nevertheless, if proper financial and administrative support existed, IPBES meetings would seem to be a natural venue for the scientific advisory bodies to collaborate.

4.4 Scientific Advisory Bodies' Responses to Outbreaks of Specific Zoonotic Diseases

This section examines to what extent scientific advisory bodies have collaborated when confronted with zoonotic diseases that relate to their respective convention's mandates. The response to highly pathogenic avian influenza offers an example of effective collaboration. The scientific advisory bodies were less engaged with respect to Ebola virus disease, while the response to COVID-19 is still emerging.

4.4.1 Highly Pathogenic Avian Influenza (HPAI)

Avian influenza implicates the health of domestic animals, wildlife, and humans. Wild waterfowl, including migratory species, are the natural reservoir for low pathogenic avian influenza (LPAI) (Cromie et al. 2012). When transmitted to domestic poultry, LPAI can mutate into HPAI, based in part on the intensiveness of farming practices (Cromie et al. 2011). HPAI is devastating to poultry populations, and it can be transmitted to wild birds and humans, primarily those who work with infected poultry. From a human health perspective, the concern is that a pandemic strain of HPAI could develop which would allow easier viral transmission from person to person (Cromie et al. 2012).

At the start of the twenty-first century, HPAI outbreaks occurred on an unprecedented scale. For example, between 2005 and 2008, HPAI H5N1 outbreaks occurred in more than 60 countries in Asia, Africa and Europe, affecting poultry, wild birds, and humans (Cromie et al. 2011). The concern that wild birds were a threat to poultry and human health led to demands in some countries that migratory birds be culled and their habitat destroyed. In response, and after discussion at meetings of the AEWA Technical Committee and CMS Scientific Council, CMS and AEWA co-convened in 2005 the Scientific Task Force on Avian Influenza and Wild Birds to provide a structure for biodiversity-related conventions and international organizations to consider the issue.

The Task Force recommended that upcoming convention meetings should take up resolutions on HPAI to emphasize that the "widespread culling of wild birds was an inappropriate response strategy" (Cromie et al. 2011). Instead, response actions should focus on biosecurity measures to ensure the separation of domestic poultry and wild birds. The Task Force's efforts led to four biodiversity-related conventions adopting resolutions on HPAI H5N1 in wild birds in a remarkably short period of time (Table 4.1). The resolutions contained similar guidance, although AEWA and CMS framed the issue on species while Ramsar focused on management of habitat.

The detailed guidance in these resolutions continues to be relevant. For example, after the spread of H1N8 in wild birds and poultry in Europe and Africa beginning in late 2016, the Ramsar STRP Chair, in a report to the Standing Committee, noted that

> "Ramsar had adopted a large body of guidance on appropriate responses to avian influenza in Resolutions IX.23 and X.21, and that this Ramsar guidance was supported by complementary guidance from the Convention on Migratory Species and the African-Eurasian Migratory Waterbirds Agreement. The [STRP] agreed that there would be value in reminding Contracting Parties of this guidance, which stresses the need to ensure that responses do not inadvertently compromise the wise use of wetlands. Furthermore, the Panel noted that lethal responses to wild birds are not supported by Ramsar guidance, or by guidance from UN-FAO and the World Organisation for Animal Health (OIE)" (Ramsar Convention 2017).

The biodiversity-related convention's experience with HPAI issues provides several lessons. First, the scientific advisory bodies, as well as the conventions, have recognized that responding to zoonotic diseases can be part of their mandate. They can also respond quickly to emerging threats, but this depends on "the

Table 4.1 Formal biodiversity-related convention Resolutions concerning HPAI H5N1 in wild birds (2005–2008)

Ramsar Convention on Wetlands	
2005	Resolution IX.23. Highly pathogenic avian influenza and its consequences for wetland and waterbird conservation and wise use
2008	Resolution X.21. Guidance on responding to the continued spread of highly pathogenic avian influenza H5N1
Convention on Migratory Species	
2005	Resolution 8.27. Migratory Species and Highly Pathogenic Avian Influenza
2008	Resolution 9.8. Responding to the challenge of emerging and re-emerging diseases in migratory species, including Highly Pathogenic Avian Influenza H5N1
African-Eurasian Waterbird Agreement	
2005	Resolution 3.18. Avian Influenza
2008	Resolution 4.15. Responding to the spread of Highly Pathogenic Avian Influenza H5N1
Convention on Biological Diversity	
2006	Decision VIII/32. Potential impact on avian influenza on biodiversity

Resolutions concerning HPAI H5N1 in wild birds (2005–2008). Source: Cromie et al. (2011). Reprinted by permission of the publisher (Taylor & Francis Ltd, http://www.tandfonline.com)

willingness of individuals to contribute their time and energy in such endeavours," as well as institutional and administrative support (Cromie et al. 2011). This institutional and administrative support must be sustained, however, and it is unclear whether it is still being provided: when clicking on the web site that was to be maintained for the Task Force (CMS 2010), one is now redirected to a page devoted to the martial arts.

4.4.2 Ebola Virus Disease (Ebola)

Ebola is a zoonotic disease, with bats believed to be a reservoir for the virus (Saéz et al. 2015). The largest Ebola outbreak, which occurred in West Africa from 2014 to 2016, afflicted more than 28,600 people and resulted in 11,325 fatalities (CDC 2020). The virus affects gorillas, chimpanzees, and duikers, and they can be a conduit for transmission to humans (CBD and WHO 2015). Accordingly, wild meat trade and consumption create risk for Ebola outbreaks. Moreover, destruction of habitat is linked to Ebola. Rulli et al. (2017) found an association between fragmented forests and Ebola outbreaks, which suggests that

> fragmented forest edges could be preferential corridors for pathogen transmission from wildlife reservoirs to humans and thereby favor the emergence of some zoonotic infections.

The response by the scientific advisory bodies has been fragmented as well, with no coordinated actions. At CBD CoP12, however, Ebola was a topic of discussion, as IUCN promoted the One Health approach, stating:

> "the CBD needs to show leadership on this issue and draw attention to the fact that it is crucial to demonstrate how habitat change and associated biodiversity loss and human health are connected. It is of fundamental importance to explain clearly how an 'upstream' approach to protect both human health and biodiversity is absolutely essential. For example, it is not widely appreciated, even amongst biodiversity experts, that land use change through deforestation is the leading driver of disease emergence in humans." (IUCN 2014).

While the CBD Secretariat has engaged through, for example, an Interagency Liaison Group with WHO (CBD 2020), SBSTTA's role appears to be more limited. SBSTTA reviewed and recommend voluntary guidance on wild meat (developed by the Secretariat), which CBD CoP14 endorsed (CBD 2018a, b). The guidance mentioned Ebola in a footnote.

At CITES CoP18, the connection between the exploitation of wildlife for meat was referenced with respect to the Center for International Forestry Research's Bushmeat Research Initiative (CITES 2019b). The Animals Committee, however, has only a tangential role regarding the wild meat trade and Ebola. The CITES CoP directed the Secretariat to review the implementation of a Resolution on wild meat and to

> "draw the attention of the Standing Committee and Animals Committee to any issues concerning international trade in wild meat" (CITES 2019c).

The issue of Ebola and biodiversity loss does not appear to be on the agenda of the other scientific advisory bodies. CMS does have an agreement on gorillas, which are at serious risk of Ebola (CMS 2019). However, neither the Scientific Council nor the Sessional Committee has been asked to address the issue.

In 2014, there was an email discussion about possible responses by the biodiversity-related conventions to the West African Ebola outbreak. Noting previous collaborations on HPAI, the Ramsar STRP Chair inquired about the extent to which Ebola was being considered by the scientific advisory bodies. While there is not a direct connection between Ebola and wetlands, it was noted that the STRP was discussing "whether there would be any value in reminding our constituents about the guidance contained in the Ramsar Wetland Disease Manual and/or the unintended consequences of habitat destruction (resulting in greater human/wildlife interaction)" (R. Gardner, personal communication, 20 October, 2014). The discussion did not lead to any coordination or collaboration among the scientific bodies.

4.4.3 COVID-19

Although COVID-19 is zoonotic in origin, the scientific advisory bodies have not made any public statements about responses or how biodiversity issues related to the coronavirus might be considered in their work. This is understandable for several reasons. It is early days in the outbreak, and the global pandemic itself has disrupted the work of the scientific advisory bodies, as meetings have either been postponed or moved on-line. Moreover, work programmes were established before the pandemic.

Nevertheless, as the responses to HPAI demonstrated, scientific advisory bodies and the biodiversity-related conventions can respond to emerging issues.

Indeed, the Secretariats of CMS and AEWA, along with the Secretariat of the Agreement on the Conservation of Populations of European Bats, quickly issued a notification entitled "Facts about Bats and COVID-19" (CMS 2020). The notification stated that "[t]here is no evidence that bats directly infected humans with COVID-19" and that "[s]cientific investigations are pointing to a chain of events that may have involved bats but most likely only through an intermediate animal." Noting the ecosystem services that bat species provide, the Secretariats criticized the killing of bats in response to COVID-19, recalling that

> [a] similar misdirected focus occurred at the height of the 2006 avian influenza, with calls for widespread culling of migratory waterbirds and the draining of their wetland habitats.

The Secretariats of CMS and AEWA also contributed to a UNEP report on zoonotic diseases issued in July 2020 (UNEP and ILRI 2020). The report discussed the coronavirus in the context of a One Health approach, explaining the relationship of land-use changes, wildlife consumption and trade, and zoonotic diseases. In a section referencing the role of CBD, CMS, and CITES, the report stated that

> [a]dditional measures to ensure safety for human health components as part of trade regulations, and improved application of all these measures at national levels, would benefit zoonotic disease control.

The CITES Secretariat, however, has actively sought to distance the convention from the issue. It expressly stated that "[m]atters regarding zoonotic diseases are outside of CITES' mandate, and therefore the CITES Secretariat does not have the competence to make comments regarding the recent news on the possible links between human consumption of wild animals and COVID-19" (CITES 2020b). Instead, it referred questions to the OIE, FAO, "and other competent bodies and research institutions at both international and national levels." By declaring these matters beyond CITES' mandate regarding trade and species conservation, the CITES Secretariat adopts a narrow view of CITES that discourages the Animals Committee's collaboration with other scientific advisory bodies.

IPBES responded to the challenge more positively, holding a workshop on biodiversity and pandemics in July 2020 (IPBES 2020c). The workshop considered the origin of zoonotic diseases, the drivers of pandemics, and policy options regarding biodiversity to prevent future outbreaks. The workshop participants were 20 experts selected by the Multidisciplinary Expert Panel from candidates nominated by IPBES members governments and scientific organizations. The workshop report, issued in October 2020, found that biodiversity loss associated with human ecological disruption contributes to pandemic risk (IPBES 2020d). Conversely, efforts to conserve protected areas and reduce unsustainable exploitation of biodiversity can assist in minimizing this risk. The report's recommendations regarding the protection of natural habitats and reduction of wildlife trade should be a clarion call for the biodiversity-related conventions and their scientific advisory bodies to promote biodiversity conservation as a preventive measure for human health.

4.5 Conclusion

There are multiple challenges for collaboration among multilateral biodiversity-related conventions. Each one has its own mandate, and some parties are keen to keep the conventions within their own lanes. Even when there is clear overlap in objectives, scientific advisory bodies must respond primarily to their core constituencies who have often approved overly ambitious work programmes that leave little room for additional activities. Moreover, effective and ongoing collaboration requires administrative support, adequate funding, and the additional time of experts who are often volunteers. Creating synergies is not a cost-free effort.

The relationship between biodiversity protection, zoonotic diseases, and human well-being is an issue that cuts across all the biodiversity-related conventions. The scientific bodies generally have the authority to respond to emerging issues, which the COVID-19 global pandemic certainly is. And the experience with HPAI demonstrates that the scientific advisory bodies can respond with guidance quickly and in a coordinated fashion.

The conventions' constituencies can ensure that the scientific advisory bodies' agendas include advice on protecting biodiversity to reduce the risks of outbreaks of infectious diseases. Such a directive could be done at a CoP. The Ramsar Convention's upcoming CoP14, scheduled to be held in Wuhan in 2021, would be an appropriate venue to make the linkage between protecting biodiversity and reducing the risk of zoonotic diseases. The Biodiversity Liaison Group or individual Standing Committees could do so as well. If the secretariats and parties are recalcitrant, observer organizations to the scientific advisory bodies can call for the inclusion of the issue on the agendas (WWF 2020).

An effort to promote biodiversity conservation as a preventive measure for human health and well-being presents an opportunity for the scientific advisory bodies to reinvigorate CSAB. An initial step could be a joint statement by CSAB to increase awareness of the connection between biodiversity conservation and disease risk reduction. CSAB could use its platform to endorse and publicize the UNEP report on zoonotic diseases. Notwithstanding the position of the CITES Secretariat, even the Chair of the Animals Committee could participate, as the Animals Committee does not report to the CITES Secretariat and the Chair is charged with giving impartial, objective scientific advice.

IPBES's foray into the issue of pandemics also provides an opportunity for CSAB. At a minimum, CSAB should endorse and support the wide dissemination of the IPBES report on biodiversity and pandemics and its key messages. Furthermore, the conventions should re-examine whether IPBES meetings (MEP or Plenary) should be used as a mechanism for CSAB to confer, if sufficient administrative support and funding is provided. If not, CSAB should consider meeting separately through electronic means. By necessity, because of the COVID-19 pandemic, we have all become more familiar and comfortable with on-line meetings. The scientific advisory bodies should take advantage of this learning experience to work together more effectively in order to achieve common biodiversity objectives.

References

AEWA. (2018a). Agreement text and annexes as amended by MOP7. Retrieved July 19, 2020, from https://www.unep-aewa.org/sites/default/files/basic_page_documents/agreement_text_english_final.pdf

AEWA. (2018b). Modus operandi of the technical committee of AEWA. Retrieved July 19, 2020, from https://www.unep-aewa.org/sites/default/files/uploads/pdf/tc_modus_operandi_approved_by_mop7_en_rev_012019.pdf

AEWA. (2018c). Resolution 7.11: Institutional arrangements: Technical committee. Retrieved July 19, 2020, from https://www.unep-aewa.org/sites/default/files/document/aewa_mop7_11_tc_en.pdf

Bowman, M., Davies, P., & Redgwell, C. (2010). *Lyster's international wildlife law*. New York: Cambridge University Press.

CBD. (2013). Report of the Sixth Meeting of the Chairs of the scientific advisory bodies of the biodiversity-related Conventions (CSAB), Formia, Italy, 7 October 2013. Retrieved July 23, 2020, from https://www.cbd.int/doc/meetings/csab/csab-06/official/csab-06-report-en.pdf

CBD. (2017). CBD SBSTTA 21/9: Guidance on integrating biodiversity considerations into one health approaches. Retrieved July 15, 2020, from https://www.cbd.int/doc/c/501c/4df1/369d06630c901cd02d4f99c7/sbstta-21-09-en.pdf

CBD. (2018a). Decision 14/4. Health and biodiversity. Retrieved July 21, 2020, from https://www.cbd.int/doc/decisions/cop-14/cop-14-dec-04-en.pdf

CBD. (2018b). Decision 14/7. Sustainable wildlife management. Retrieved July 26, 2020, from https://www.cbd.int/doc/decisions/cop-14/cop-14-dec-07-en.pdf

CBD. (2020). Interagency Liaison Group on biodiversity and health. Retrieved July 26, 2020, from https://www.cbd.int/health/ilg-health/#tab=1

CBD and WHO. (2015). Connecting global priorities: Biodiversity and human health, a state of knowledge review. Retrieved July 26, 2020, from https://www.cbd.int/health/summary-state-knowledge-review-en.pdf

CDC. (2020). 2014-2016 Ebola outbreak in West Africa. Retrieved July 26, 2020, from https://www.cdc.gov/vhf/ebola/history/2014-2016-outbreak/index.html

CITES. (2012a). AC26 Doc. 23 (Rev.1): Relationship between wildlife trade and wildlife diseases. Retrieved July 17, 2020, from https://www.cites.org/sites/default/files/eng/com/ac/26/E26-23.pdf

CITES. (2012b). AC26: Summary record. Retrieved July 17, 2020, from https://www.cites.org/sites/default/files/eng/com/ac/26/sum/E-AC26-SumRec.pdf

CITES. (2018a). Rules of procedure for meetings of the animals committee. Retrieved July 17, 2020, from https://cites.org/sites/default/files/eng/com/ac/E-AC30-RoP-2018.pdf

CITES. (2018b). Thirtieth meeting of the animals committee: List of participants. Retrieved July 17, 2020, from https://cites.org/sites/default/files/eng/com/AC/30/AC30ParticipantListFINAL.pdf

CITES. (2019a). Conf. 18.2 establishment of committees. Retrieved July 17, 2020, from https://cites.org/sites/default/files/document/E-Res-18-02.pdf

CITES. (2019b). COP18 Doc. 95: Guidance materials, activities and tools aimed at enhancing parties' capacity to regulate bushmeat trade. Retrieved July 17, 2020, from https://cites.org/sites/default/files/eng/cop/18/doc/E-CoP18-095.pdf

CITES. (2019c). Conf. 13.11 (Rev. CoP18) Wild meat. Retrieved July 26, 2020, from https://www.cites.org/sites/default/files/document/E-Res-13-11-R18.pdf

CITES. (2020a). Animals and Plants Committees. Retrieved July 17, 2020, from https://cites.org/eng/disc/ac_pc.php

CITES. (2020b). CITES Secretariat's statement in relation to COVID-19. Retrieved July 17, 2020, from https://cites.org/eng/CITES_Secretariat_statement_in_relation_to_COVID19

CITES and OIE. (2015). Cooperation agreement between the CITES Secretariat and the OIE. Retrieved July 17, 2020, from https://www.oie.int/fileadmin/Home/eng/About_us/docs/pdf/accords/Accord_CITES_OIE_novembre_2015.pdf

CMS. (1979). Retrieved July 18, 2020, from https://www.cms.int/sites/default/files/instrument/CMS-text.en_.PDF

CMS. (2008). Resolution 9.8: Responding to the challenges of emerging and re-emerging diseases in migratory species, including highly pathogenic Avian Influenza H5N1. Retrieved July 31, 2020, from https://www.cms.int/sites/default/files/document/Res_9_08_Wildlife_Disease_En.pdf

CMS. (2010). UNEP/CMS/ScC16/Inf.7: Conclusions and recommendations of the Third Technical Meeting of the scientific task force on Avian Influenza and wild birds: HPAI H5N1 and wild birds: Reviewing the global issue and assessing future priorities. Retrieved July 25, 2020, from https://www.cms.int/sites/default/files/document/ScC16_Inf_07_Scientific_Task_Force_on_AI_Eonly_0.pdf

CMS. (2014). UNEP/CMS/ScC18/Doc.4.4: Options for a new structure of the scientific council. Retrieved July 18, 2020, from https://www.cms.int/sites/default/files/document/Doc_4_4_Restructuring%20of%20ScC_E.pdf

CMS. (2017). Resolution 12.4: Scientific council. Retrieved July 18, 2020, from https://www.cms.int/sites/default/files/document/cms_cop12_res.12.4_scientific-council_e.pdf

CMS. (2019). UNEP/GA/MOP3/REPORT: Agreement on the conservation of Gorillas and their habitats of the convention on migratory species, report of the third meeting of the parties. Retrieved July 26, 2020, from https://www.cms.int/gorilla/sites/default/files/document/cms_ga-mop3_report_without-annexes_e_1.pdf

CMS. (2020). 2020/005: Facts about bats and COVID-19. Retrieved July 27, 2020, from https://www.cms.int/en/news/2020005-facts-about-bats-and-covid-19

Cook, R. A., Karesh, W. B., & Osofsky, S. A. (2004). The Manhattan principles on "One World, One Health." Retrieved July 15, 2020, from http://www.oneworldonehealth.org/sept2004/owoh_sept04.html

Cromie, R., Davidson, N., Galbraith, C., Hagemeijer, W., Horwitz, P., Lee, R., et al. (2011). Responding to emerging challenges: Multilateral environmental agreements and highly pathogenic avian influenza H5N1. *Journal of International Wildlife Law & Policy, 14*, 206–242.

Cromie, R. L., Lee, R., Delahay, R. J., Newth, J. L., O'Brien, M. F., Fairlamb, H. A., et al. (2012). *Ramsar wetland disease manual: Guidelines for assessment, monitoring and management of animal disease in wetlands. Ramsar Technical Report No. 7*. Gland: Ramsar Convention Secretariat. Retrieved July 16, 2020, from https://www.ramsar.org/sites/default/files/documents/library/rtr7-disease.pdf

de Groot, R. S., Stuip, M. A. M., Finlayson, C. M., & Davidson, N. (2006). *Valuing wetlands: Guidance for valuing the benefits derived from wetland ecosystem services, Ramsar Technical Report No. 3/CBD Technical Series No. 27*. Gland/Montreal: Ramsar Convention Secretariat/Secretariat of the Convention on Biological Diversity. Retrieved July 23, 2020, from https://www.cbd.int/doc/publications/cbd-ts-27.pdf

Díaz, S., Demissew, S., Carabias, J., Joly, C., Lonsdale, M., Ash, N., et al. (2015). The IPBES conceptual framework – Connecting nature and people. *Current Opinion in Environmental Sustainability, 14*, 1–16.

Favre, D. S. (1989). *International trade in endangered species: A guide to CITES*. Dordrecht: Kluwer Academic Publishers.

Finlayson, C. M., & Davidson, N. C. (2018). *Global wetland outlook: Technical note on status and trends*. Gland: Ramsar Convention Secretariat. Retrieved July 16, 2020, from https://static1.squarespace.com/static/5b256c78e17ba335ea89fe1f/t/5bcaeb0624a694d6076938eb/1540025105566/TN+-+Status+final+051018.pdf

Gardner, R. C., & Davidson, N. C. (2011). The Ramsar Convention. In B. A. LePage (Ed.), *Wetlands: Integrating multidisciplinary concepts* (pp. 189–203). Dordrecht: Springer.

Gardner, R. C., & Finlayson, C. M. (2018). *Global wetland outlook; State of the world's wetlands and their ecosystem services*. Gland: Ramsar Convention.

Gardner, R. C., & Grobicki, A. (2016). Synergies between the Convention on Wetlands of International Importance, especially as Waterfowl Habitat and other multilateral agreements: Possibilities and pitfalls. In *UN Environment, understanding synergies and mainstreaming among the biodiversity related conventions: A special contributory volume by key convention secretariats and scientific bodies* (pp. 54–67). Nairobi: UN Environment.

Gitay, H., Finlayson, C. M., & Davidson, N. C. (2011). *A Framework for assessing the vulnerability of wetlands to climate change. Ramsar Technical Report No. 5/CBD Technical Series No. 57*. Gland/Montreal: Ramsar Convention Secretariat/Secretariat of the Convention on Biological Diversity. Retrieved July 23, 2020, from https://www.ramsar.org/sites/default/files/documents/pdf/lib/lib_rtr05.pdf

Horwitz, P., Finlayson, M., & Weinstein, P. (2012). *Healthy wetlands, healthy people: A review of wetlands and human health interactions. Ramsar Technical Report No. 6*. Gland/Geneva: Secretariat of the Ramsar Convention on Wetlands/The World Health Organization. https://www.ramsar.org/sites/default/files/documents/pdf/lib/rtr6-health.pdf

Hotes, S., & Opgenoorth, L. (2014). Trust and control at the science–policy interface in IPBES. *BioScience, 64*(4), 277–278. https://doi.org/10.1093/biosci/biu019

Intergovernmental Science–Policy Platform on Biodiversity and Ecosystem Services (IPBES). (2019). *Summary for policymakers of the global assessment report on biodiversity and ecosystem services*. Bonn: IPBES Secretariat.

IPBES. (2020a). Engagement related to multilateral environmental agreements. Retrieved July 24, 2020, from https://ipbes.net/multilateral-environmental-agreements

IPBES. (2020b). Rules of procedure for sessions of the Plenary of the Intergovernmental Science Policy Platform on Biodiversity and Ecosystem Services. Retrieved July 24, 2020, from https://ipbes.net/sites/default/files/downloads/IPBES_rules_of_procedure.pdf

IPBES. (2020c). IPBES workshop on biodiversity and pandemics. Retrieved July 27, 2020, from https://ipbes.net/pandemics-workshop

IPBES. (2020d). Workshop report on biodiversity and pandemics of the intergovernmental platform on biodiversity and ecosystem services. Retrieved July 20, 2020, from https://ipbes.net/sites/default/files/2020-12/IPBES%20Workshop%20on%20Biodiversity%20and%20Pandemics%20Report_0.pdf

IUCN. (2014). Ebola outbreak highlights critical links between biodiversity loss and human health says IUCN's Wildlife Health Specialist Group. Retrieved July 31, 2020, from https://www.iucn.org/content/ebola-outbreak-highlights-critical-links-between-biodiversity-loss-and-human-health-says-iucns-wildlife-health-specialist-group

Jones, K. E., Patel, N. G., Levy, M. A., Storeygard, A., Balk, D., Gittleman, J. L., et al. (2008). Global trends in emerging infectious diseases. *Nature, 451*, 990–993.

Lewis, M. (2016). AEWA at twenty: An appraisal of the African-Eurasian waterbird agreement and its unique place in International Environmental Law. *Journal of International Wildlife Law & Policy, 19*(1), 22–61.

OIE. (2020). Working group on wildlife. Retrieved July 17, 2020, from https://www.oie.int/en/standard-setting/specialists-commissions-working-ad-hoc-groups/working-groups-reports/working-group-on-wildlife/

Ramsar Convention. (1993). Resolution 5.5: Establishment of a scientific and technical review panel. Retrieved July 16, 2020, from https://www.ramsar.org/sites/default/files/documents/pdf/res/key_res_5.5e.pdf

Ramsar Convention. (1996). Resolution VI.7: The scientific and technical review panel. Retrieved July 16, 2020, from https://www.ramsar.org/sites/default/files/documents/pdf/res/key_res_vi.07e.pdf

Ramsar Convention. (2012). Resolution XI.12: Wetlands and health: Taking an ecosystem approach. Retrieved July 16, 2020, from https://www.ramsar.org/sites/default/files/documents/pdf/cop11/res/cop11-res12-e.pdf

Ramsar Convention. (2015). Resolution XII.5: New framework for delivery of scientific and technical advice and guidance on the Convention. Retrieved July 16, 2020, from https://www.ramsar.org/sites/default/files/documents/library/cop12_res05_new_strp_e_0.pdf

Ramsar Convention. (2016a). SC52-10: Report of the Chair of the Scientific and Technical Review Panel (STRP) and amended draft STRP 2016-2018 work plan. Retrieved July 16, 2020, from https://www.ramsar.org/sites/default/files/documents/library/sc52-10_strp_chair_report_e.pdf

Ramsar Convention. (2016b). Report and decisions of the 52nd Meeting of the Standing Committee. Retrieved July 16, 2020, from https://www.ramsar.org/sites/default/files/documents/library/sc52_report_and_decisions_e.pdf

Ramsar Convention. (2017). SC53-10: Report of the Chair of the Scientific and Technical Review Panel (STRP). Retrieved July 25, 2020, from https://www.ramsar.org/sites/default/files/documents/library/sc53-10_strp_chair_report_e.pdf

Rulli, M. C., Santini, M., Hayman, D. T. S., & D'Odorico, P. (2017). The nexus between forest fragmentation in Africa and Ebola virus disease outbreaks. *Scientific Reports, 7*, 41613. https://doi.org/10.1038/srep41613

Saéz, A. M., Weiss, S., Nowak, K., Lapeyre, V., Zimmermann, F., Düx, A., et al. (2015). Investigating the zoonotic origin of the West African Ebola epidemic. *EMBO Molecular Medicine, 7*, 17–23.

Secretariat of the CBD. (2005). *Handbook of the convention on biological diversity including its Cartagena protocol on biosafety* (3rd ed.). Canada: Montreal. Retrieved July 21, 2020, from https://www.cbd.int/doc/handbook/cbd-hb-all-en.pdf

UNEP. (2015). *Sourcebook of opportunities for enhancing cooperation among Biodiversity-related Conventions at national and regional levels*. Nairobi: UNEP.

UNEP and ILRI. (2020). *Preventing the Next Pandemic: Zoonotic diseases and how to break the chain of transmission*. Nairobi: Nairobi. Retrieved July 27, 2020, from https://wedocs.unep.org/bitstream/handle/20.500.11822/32316/ZP.pdf?sequence=1&isAllowed=y

UNEP and WCMC. (2012). Promoting synergies within the cluster of biodiversity-related multilateral environmental agreements. Retrieved July 23, 2020, from https://www.cbd.int/doc/nr/Promoting_synergies_in_the_biodiversity_cluster.pdf

United Nations (UN). (1992). Convention on biological diversity. Retrieved July 21, 2020, from https://www.cbd.int/doc/legal/cbd-en.pdf

United Nations (UN). (2015). *Transforming our world: The 2030 agenda for sustainable development, A/RES/70/1*. New York: UN General Assembly. Retrieved July 15, 2020, from https://www.un.org/en/development/desa/population/migration/generalassembly/docs/globalcompact/A_RES_70_1_E.pdf

Urho, N. (2009). *Possibilities of enhancing co-operation and co-ordination among MEAs in the biodiversity cluster*. Copenhagen: Nordic Council of Ministers. Retrieved July 20, 2020, from https://www.cbd.int/doc/meetings/biodiv/brcws-2016-01/other/brcws-2016-01-erho-en.pdf

Van Vliet, N., Gomez, J., & Restrepo, S. (2016). Bushmeat, CITES and livelihoods. In *Handbook for assessing the impacts of implementing CITES regulations for bushmeat species on the livelihoods of rural communities*. Bogor: Center for International Forestry Research. Retrieved July 17, 2020, from https://www.cifor.org/knowledge/publication/5336/

WHO. (2020). Coronavirus disease 2019 (COVID-19) situation report – 94. Retrieved July 15, 2020, from https://www.who.int/docs/default-source/coronaviruse/situation-reports/20200423-sitrep-94-covid-19.pdf

WHO and Secretariat of the CBD. (2020). Biodiversity & infectious diseases questions & answers. Retrieved July 15, 2020, from https://www.who.int/docs/default-source/climate-change/qa-infectiousdiseases-who.pdf?sfvrsn=3a624917_3

Wolfe, N. D., Daszak, P., Kilpatrick, A. M., & Burke, D. S. (2005). Bushmeat hunting and deforestation in prediction of zoonoses emergence. *Emerging Infectious Diseases, 11*, 1822–1827. https://doi.org/10.3201/eid1112.040789

WWF. (2020). COVID-19 and wildlife trade: Perspectives and proposed actions. Retrieved July 27, 2020, from https://www.worldwildlife.org/pages/covid-19-and-wildlife-trade-perspectives-and-proposed-actions

Part II
Biodiversity Interactions with Global Conventional Regimes

Chapter 5
Conservation and Sustainable Use of Biodiversity in the International Regulation of Trade in Goods

Xavier Fernández-Pons

Contents

5.1 Introduction 80
5.2 Legal Bases of the Multilateral Trading System 81
  5.2.1 The GATT 1947 System 81
  5.2.2 The Marrakesh Agreements 83
5.3 Cases in the WTO Dispute Settlement System 84
  5.3.1 The US: Shrimp Case 85
  5.3.2 The US: Tuna II (Mexico) Case 86
  5.3.3 Other Cases 88
5.4 Multilateral and Plurilateral Negotiations in the WTO 90
5.5 Provisions in Regional Trade Agreements 92
5.6 Conclusion 94
References 96

Abstract This chapter examines how the objectives of conservation and sustainable use of biodiversity have been considered within the rules of the multilateral trading system and other international agreements on trade in goods. Initially, the concept of development was conceived in productivist terms, without any explicit reference to the protection of biodiversity. The Marrakesh Agreements incorporated the objective of sustainable development but introduced few environmental provisions in their precepts. The dispute settlement system of the World Trade Organization (WTO) has been highly relevant for the promotion of conciliation between trade and the environment, interpreting criteria to determine whether trade restrictions imposed for the conservation and sustainable use of biodiversity are compatible with WTO rules. However, an analysis of such case law shows that this interpretative path has limitations, and it underlines the need for more ambitious negotiations. The stalled Doha Round is now focusing its efforts on the adoption of a new multilateral agreement against the most harmful fisheries subsidies, which is explicitly linked to the Sustainable Development Goals (SDGs). New regional trade agreements are

X. Fernández-Pons (✉)
Universitat de Barcelona, Faculty of Law, Barcelona, Spain
e-mail: xavierfernandez@ub.edu

M. Campins Eritja, T. Fajardo del Castillo (eds.), *Biological Diversity and International Law*, https://doi.org/10.1007/978-3-030-72961-5_5

incorporating many provisions on sustainable development and trade in forest products, fisheries and wildlife protection. These trends towards green trade agreements are promising, but the effective application of such systemic interconnections is still a major challenge.

5.1 Introduction

The relationships between global trade liberalization and biodiversity protection are multiple and complex. The logic of trade liberalization tends to promote large-scale production, the establishment of global value chains and the uniformization of certain consumption patterns across the planet. In contrast, the logic of biodiversity protection is to preserve the variability among ecosystems, species and genetic resources or, in other words, "the diversity of life forms on earth" (Deke 2008, p. 1).

Trying to reconcile the two logics is not easy. The notion of sustainable development, which combines economic, social and environmental aspects, seeks the conciliation between economic development and environmental protection. The Convention on Biological Diversity (CBD) takes up the notion of sustainable development by including among its objectives, which are set out in Art. 1, the conservation of biological diversity and "the sustainable use of its components", an expression that is then defined in Art. 2 as "the use of components of biological diversity in a way and at a rate that does not lead to the long-term decline of biological diversity, thereby maintaining its potential to meet the needs and aspirations of present and future generations". In particular, the sustainable use of biodiversity calls for "sustainable trade" (Costanza et al. 1995, p. 16), which subjects the exchange of goods to biodiversity requirements, taking into account the environmental conditions of production and the impact of "processes and production methods" (PPMs) on biodiversity (OECD 1994; Holzer 2017).

Trade restrictions and other trade-related measures (e.g. taxes, certifications and labels) on goods adopted by states or by the European Union (EU) to protect biodiversity can be classified from two main points of view:

(a) On the one hand, depending on their legal basis, they are measures adopted in application of multilateral environmental agreements (MEAs) which, in certain cases, include trade provisions (such as the Convention on International Trade in Endangered Species of Wild Fauna and Flora, CITES) or they are measures adopted unilaterally by one state (or by the EU) based on its own internal regulations (such as the prohibition of certain pesticides harmful to bees and other pollinators in the EU).
(b) On the other hand, depending on their object, such measures may relate to a product as such (e.g. measures to prevent the introduction of alien species that threaten ecosystems, habitats or native species) or to PPMs, which may in turn be of two subtypes, product-related PPMs (which leave traces in the final

product, such as a regulation on the maximum residue level of pesticides) or non-product-related PPMs (which affect the environment without being reflected in the physical characteristics of the final product, such as sustainable timber certifications).

When a state (or the EU) adopts such measures, it is sometimes questioned whether they are compatible with the rules of the World Trade Organization (WTO) or other international trade agreements. Trade measures based on an MEA do not usually generate disputes, particularly if all the states concerned have consented to the agreement. The most controversial measures tend to be unilateral ones relating to non-product-related PPMs, which are sometimes perceived, particularly by developing countries, as forms of green protectionism or extraterritorial interference in the management of natural resources under their respective national jurisdictions and which deserve extensive analysis among legal scholars (Conrad 2011; Cooreman 2017; Maggio 2017; Sifonios 2018). It should be remembered that according to Principle 12 of the Rio Declaration "unilateral actions to deal with environmental challenges outside the jurisdiction of the importing country should be avoided" and "environmental measures addressing transboundary or global environmental problems should, as far as possible, be based on an international consensus".

This chapter attempts to analyse the extent to which the objectives of conservation and sustainable use of biodiversity have been assumed within the rules of the multilateral trading system and other international agreements on trade in goods. To this end, the chapter begins in Sect. 5.2 by recalling the origins of the issue under the system of the 1947 General Agreement on Tariffs and Trade (GATT) and the provisions of the 1994 Marrakesh Agreements. Then Sect. 5.3 examines various cases raised through the WTO dispute settlement system regarding trade restrictions (or other trade measures) on goods to protect biodiversity, observing the contributions (and limitations) of its jurisprudence. This is followed by Sect. 5.4, which assesses the trade negotiations conducted under the WTO, including current proposals for the prohibition of the most harmful subsidies in the fishing sector, which are explicitly linked to Target 14.6 of the Sustainable Development Goals (SDGs); Sect. 5.5, which examines the growing inclusion of environmental (and biodiversity) provisions in regional trade agreements, such as those concluded by the EU with third countries, which have recently tended to include a chapter on "trade and sustainable development"; and lastly Sect. 5.6, which puts forward conclusions on the positive and negative aspects surrounding the issue.

5.2 Legal Bases of the Multilateral Trading System

5.2.1 The GATT 1947 System

At the outset of the multilateral trading system, with GATT 1947, environmental issues were not high on the international agenda. The preamble to GATT 1947 reflected a productivist conception of development that was prevalent at the time, stating that international trade should tend towards "developing the full use of the resources of the world and expanding the production and exchange of goods".

The text of GATT 1947 did not include any explicit mention of the environment. Its basic principles focused on market access and non-discrimination between like products, and the analysis of likeness mainly centred on the physical characteristics of products and their commercial uses, regardless of possible distinctions based on non-product-related PPMs (GATT 1970, para. 18). In its exceptions, it recognized that the contracting parties could impose trade restrictions (or other trade measures) to preserve certain legitimate objectives, which, while not explicitly referring to environmental protection, could justify certain measures taken for environmental reasons. Thus, Art. XX of GATT 1947, on "general exceptions", justified: (b) measures "necessary to protect human, animal or plant life or health"; and (g) measures "relating to the conservation of exhaustible natural resources if such measures are made effective in conjunction with restrictions on domestic production and consumption". The introductory clause (*chapeau*) of Art. XX also required that such measures should not be applied

> in a manner which would constitute a means of arbitrary or unjustifiable discrimination between countries where the same conditions prevail, or a disguised restriction on international trade.

In 1971, coinciding with the preparation of the UN Conference on the Human Environment held in Stockholm in 1972, the Director-General of GATT 1947 established a group on environmental measures and international trade (EMIT), but the group's activities were limited (Nordström and Vaughan 1999, pp. 67–72). The Tokyo Round (1973–1979) adopted a first Agreement on Technical Barriers to Trade (Tokyo TBT), which recognized that "technical specifications" of products could be adopted, among other legitimate objectives, for the protection of the "environment" (Art. 2.2), but clarified that technical specifications referred to "characteristics of a product such as levels of quality, performance, safety or dimensions" (Annex 1.1), without referring to PPMs.

In the early nineties, a controversial panel report on the *United States (US)—Restrictions on Imports of Tuna (Mexico)* case, circulated on 3 September 1991, highlighted the reluctance of the GATT 1947 system to justify unilateral trade restrictions for the conservation of biodiversity based on non-product-related PPMs (GATT 1991). On 5 November 1990, Mexico requested consultations with the US concerning restrictions on imports of tuna and tuna products that invoked the protection of dolphins. The US observed that in certain areas of the world, particularly in the Eastern Tropical Pacific Ocean, there is a close association between tuna and dolphins. As a result, when tuna is harvested with purse seine nets, numerous dolphins may be killed incidentally. The US imposed on vessels under its jurisdiction the use of certain procedures to eliminate or reduce such incidental kills of dolphins. At the same time, the US established a prohibition on imports of tuna and tuna products from other countries that did not adopt a comparable regime, which especially affected Mexico.

In its claim, Mexico asserted that the restrictions were incompatible with a number of basic principles of GATT 1947. The US sought to justify its restrictions by invoking the exceptions contained in Art. XX. The panel largely embraced Mexico's arguments and gave considerable weight to the historical interpretation of GATT 1947, repeatedly referring to its *travaux préparatoires*. The panel concluded that the US import restrictions were inconsistent with basic principles and could not be justified under Art. XX(b), observing that the drafters of the provision "focused on the use of sanitary measures to safeguard life or health of humans, animals or plants within the jurisdiction of the importing country" (GATT 1991, para. 5.26), nor could they be justified under Art. XX(g), which "was intended to permit contracting parties to take trade measures primarily aimed at rendering effective restrictions on production or consumption within their jurisdiction", rejecting the "extrajurisdictional interpretation" suggested by the US to conserve exhaustible natural resources abroad (GATT 1991, paras. 5.31 and 5.32). According to the panel, extrajurisdictional environmental problems had to be addressed through "the negotiation of international cooperative arrangements" (GATT 1991, para. 5.28), leaving unanswered what to do if such negotiations proved unsuccessful.

The US was very disappointed with the report and another successive report in a very similar case brought by the European Economic Community, called *US – Tuna (EEC)* (GATT 1994, para. 6.1). Neither report was adopted. Criticism of the *Tuna–Dolphin* cases abounded among environmentalists, who saw them as a clear example of the GATT 1947 system's lack of sensitivity to growing global environmental challenges (Hurlock 1992; McDorman 1992) and underlined the need to green the multilateral trading system (Esty 1993).

5.2.2 The Marrakesh Agreements

The Marrakesh Agreements, signed on 15 April 1994 and entering into force on 1 January 1995, were influenced by the critics mentioned above and by the momentous United Nations Conference on Environment and Development (UNCED) held in Rio in 1992. The preamble to the Marrakesh Agreement establishing the WTO refers to the "optimal use of the world's resources in accordance with the objective of sustainable development [...] to protect and preserve the environment". However, references to the environment were very scarce in the remainder of the Marrakesh Agreements.

For example, GATT 1947 was formally replaced by GATT 1994, but the text remained almost unchanged. It is striking that the wording of the general exceptions contained in Art. XX remained the same, without incorporating any explicit reference to the environment, biodiversity or climate change or clarifying the thorny issue of the jurisdictional scope of its various paragraphs.

The preamble of the Agreement on Agriculture, devoted to establish "a fair and market-oriented agricultural trading system", made some references to non-trade concerns, including "the need to protect the environment", but its precepts do little to regulate this question.

Although this chapter focuses on international trade in goods, it must be noted that the General Agreement on Trade in Services (GATS) included a provision on general exceptions in its Art. XIV, which was limited, essentially, to reproducing the provisions of the classic Art. XX of GATT, with the striking absence of a paragraph on measures relating to the conservation of exhaustible natural resources.

The Agreement on Trade-Related Aspects of Intellectual Property Rights (TRIPs) recognized, in Art. 27.2, that members "may exclude" from patentability inventions "to protect human, animal or plant life or health or to avoid serious prejudice to the environment". Art. 27.3(b) provided that members "may also exclude" from patentability "plants and animals other than micro-organisms, and essentially biological processes for the production of plants or animals other than non-biological and microbiological processes". The patentability of genetic resources contemplated under TRIPs has led to a heated debate on the extent to which it is compatible with the objective of the CBD on "the fair and equitable sharing of the benefits arising out of the utilization of genetic resources" (Art. 1 of CBD) and the protection of the "knowledge, innovations and practices of indigenous and local communities" (Art. 8 (j) of CBD), developed in the Nagoya Protocol. In any case, these questions lie outside the specific scope of this chapter.

The limited references to environmental issues in the Marrakesh Agreements were to some extent offset by the Marrakesh Ministerial Decision on Trade and Environment, which led to the creation of the WTO Committee on Trade and Environment (CTE). The CTE has a broad agenda and engages in valuable discussions that seek to reconcile international trade with environmental protection, addressing issues such as the relationship between the WTO and MEAs, charges and taxes for environmental purposes, exports of domestically prohibited goods and eco-labelling (Sinha 2013). However, the CTE does not have any decision-making powers and the legal results of its work depend on negotiations and consensus among WTO members, which, as will be shown in greater detail below, have so far produced few legal results.

Thus, it is understandable that the WTO dispute settlement system and its adjudicative bodies (panels and the Appellate Body (AB)) have been playing an important role in interpreting the compatibility of certain trade-related environmental measures with current WTO rules.

5.3 Cases in the WTO Dispute Settlement System

In the WTO dispute settlement system, there have been several cases on trade and the environment that specifically concern certain living organisms. This chapter will not go into cases that focus primarily on sanitary or phytosanitary arguments (such as the *European Communities (EC)—Hormones* case and the *EC—Approval and Marketing of Biotech Products* case) or on animal welfare and public morals (such as the

EC—Seal Products case). Rather, it will focus on cases that revolve around the conservation and sustainable use of biodiversity. The most significant dispute on this issue is the *US—Shrimp* case, which has been widely analysed by scholars and is considered a systemic milestone and even a legal "revolution" in the interpretation of the multilateral trade rules (Yavitz 2001; Howse 2002; Juste Ruiz 2003; Fernández Egea 2008).

5.3.1 The US: Shrimp Case

The facts of the case are very similar to those in the *Tuna–Dolphin* cases under GATT 1947. The case, which is also known as the *Shrimp–Turtle* case, refers to a unilateral trade restriction based on non-product-related PPMs. The complaining parties were four Asian countries (India, Malaysia, Pakistan and Thailand) which, on 8 October 1996, requested consultations with the US concerning a ban on the importation of shrimp and shrimp products (WTO 1996). The US noted that shrimps and marine turtles often coexist in certain waters and that shrimp fishing with nets usually results in the accidental death of numerous marine turtles, including endangered species. Within its jurisdiction, the US imposed the use of filters called turtle excluder devices (TEDs) to prevent such turtle bycatch. The US also noted that TEDs were not used in other countries and decided to prohibit the import of shrimp from them. The complaining countries claimed that the ban was contrary to basic principles of the current GATT 1994. The US based its defence mainly on the general exception in Art. XX(g) of GATT 1994, on the grounds that the ban was a measure "relating to the conservation of exhaustible natural resources".

The AB, in its 1998 report, accepted many of the US's arguments (WTO 1998). It noted that the provisions of GATT 1994 must be interpreted in light of the preamble to the Marrakesh Agreement establishing the WTO, which expressly refers, as we have seen, to the "objective of sustainable development" (WTO 1998, para. 129). The AB interpreted that the notion of "exhaustible natural resources" not only refers, as the claiming parties intended, to non-renewable inert resources, but can also cover living and renewable resources, such as marine turtles, explicitly citing the Agenda 21 and other international environmental instruments, such as the CBD (WTO 1998, paras. 130–132). It also considered that the US was permitted to impose a trade restriction to conserve natural resources beyond its jurisdiction, on the understanding that Art. XX(g) can cover measures with an extraterritorial scope as long as "there is a sufficient nexus", such as the highly migratory nature of marine turtles (WTO 1998, para. 133). Notwithstanding all these valuable interpretations, the AB concluded that the way in which the US had applied the ban did not comply with the requirements of the *chapeau* of Art. XX, as the ban had a very inflexible design (requiring the use of TEDs such as those used in the US) and it was adopted without any prior cooperative effort with the Asian countries, which contrasted with the negotiations carried out with other interested countries, such as those in the Caribbean (WTO 1998, paras. 165–186). This interpretation of the *chapeau* confirmed the preference to address extraterritorial environmental problems through cooperation and MEAs, in line with the aforementioned Principle 12 of the Rio Declaration.

Following the report, the US improved various aspects of its measure. However, Malaysia refused to cooperate and continued to allow shrimp fishing techniques that resulted in the accidental death of marine turtles. The US lifted the ban on the import of shrimps from India, Pakistan and Thailand, but kept the ban in place for Malaysia. On 12 October 2000, Malaysia alleged that the US had not implemented the adopted reports appropriately and requested, pursuant to the compliance proceedings of Art. 21.5 of the Dispute Settlement Understanding (DSU), that the matter should be referred again to the original panel (WTO 2000a). Regarding the claim of Malaysia, both the panel (WTO 2001a) and the AB (WTO 2001b) concluded that the continuing US ban on the import of shrimps from Malaysia was fully compatible with GATT 1994, because it was covered under paragraph (g) of Art. XX and now respected the conditions of its *chapeau.*

In sum, the *US—Shrimp* case showed that WTO law can admit trade restrictions that are based on non-product-related PPMs and adopted unilaterally by a member for the conservation of exhaustible natural resources abroad, if there is a "sufficient nexus" and it has previously made serious good faith efforts at cooperation (Charnovitz 2002). However, the case dealt with a very specific situation and it is not easy to extrapolate to other possible situations. For example, the extraterritoriality of the US measure was justified in this case by the highly migratory nature of marine turtles in the oceans, but what would happen if a member wanted to adopt trade restrictions to protect a species of turtle (or other animal or plant) that lived exclusively in an inland lake in another country? Thinking about this and other hypotheses, Young (2014) notes that there are still many uncertainties in the justification of certain extraterritorial measures under WTO rules.

5.3.2 The US: Tuna II (Mexico) Case

Among the disputes that focus on measures relating to the conservation of biodiversity, the *US—Tuna II (Mexico)* case stands out as well. The case connects with the *Tuna–Dolphin* disputes under the old GATT 1947 system. In this case, however, the complaint brought by Mexico, which requested consultations with the US on 24 October 2008, refers exclusively to the "dolphin-safe" labelling scheme established by US authorities, which is clearly based on non-product-related PPMs, informing consumers that the tuna has been caught by methods that prevent the accidental death of dolphins (WTO 2008). Mexico did not challenge the existence of such labelling *per se*. Rather, Mexico complained about why the US was demanding more requirements for the labelling of tuna caught in the Eastern Tropical Pacific Ocean (ETP), where most Mexican tuna fishermen usually fish, than for labelling tuna caught in other ocean areas. The US based the distinction on the fact that there is a special association between tuna and dolphins in the ETP.

However, Mexico disagreed, arguing that there are similar situations and risks in other ocean areas.

From the perspective of this chapter, it is interesting to note that the reports on this case from the panel (WTO 2011) and the AB (WTO 2012) examined the "dolphin-safe" label under the current TBT Agreement (TBT). For a long time, some scholars have discussed whether the definitions of "technical regulation" and "standard" in Annex 1 of the TBT cover a labelling scheme on non-product-related PPMs, such as "dolphin-safe" labelling (Fernández Pons and Baroncini 2004, pp. 142–146). In this case, however, the parties accepted that a label such as "dolphin-safe" was covered by the TBT. The panel and the AB did not question the concurrent views of the two parties on this point. Thus, they endorsed interpretations in favour of considering labelling requirements relating to non-product-related PPMs as a possible type of technical specification covered under the TBT. This opens the way to argue for the compatibility with WTO rules of other possible labelling requirements for non-product-related PPMs, such as those relating to wood or paper obtained from sustainable sources or showing the ecological footprint of products (Ankersmit and Lawrence 2012; Appleton 2015).

However, the fact that a label on non-product-related PPMs is covered under the TBT's definitions does not mean, obviously, that it is now fully justified, since such a label still has to comply with the various provisions of the TBT, such as those relating to non-discrimination, transparency and harmonization. Thus, in this case, the AB concluded, in its 2012 report, that "dolphin-safe" labelling was incompatible with Art. 2.1 of the TBT, as the higher level of requirements for tuna fishing in the ETP in comparison to other ocean areas was not well founded (WTO 2012, paras. 298, 299 and 407).

In theory, the US could have chosen to relax the specific requirements for the ETP, but instead decided in successive reforms of its "dolphin-safe" labelling scheme (in 2013 and 2016) to increase the requirements for other ocean areas. Finally, after successive reports on compliance, the AB concluded, in a report circulated on 14 December 2018 and adopted on 11 January 2019, that the current US "dolphin-safe" labelling scheme is now fully compatible with WTO rules, since the 2016 tuna measure, as a whole, is based on a "legitimate regulatory distinction"; is "calibrated to the risks to dolphins arising from the use of different fishing methods in different areas of the ocean"; and does not commit "arbitrary or unjustifiable discrimination" (WTO 2018, paras. 7.2, 7.10 and 7.14). Some commentators celebrate that the end of this long dispute has led to increasing and extending the requirements for the protection of dolphins and consider that the current US "dolphin-safe" labelling scheme should be clearly qualified "as a sustainability discipline" (Baroncini and Brunel 2020, p. 205).

5.3.3 Other Cases

Other cases about trade restrictions to promote the sustainable use of biodiversity have been resolved through mutually agreed solutions. This is the case with the *Chile—Swordfish* dispute. The EU filed a complaint against Chile at the WTO on 19 April 2000, alleging that Chile had prohibited the landing of swordfish caught by EU vessels in the international waters of the South-Eastern Pacific Ocean and that this was incompatible with Art. V of GATT 1994, which provides for "freedom of transit" (WTO 2000b). Chile sought to justify its restrictions before the WTO under Art. XX(g) of GATT 1994, arguing that the measures related to the conservation of exhaustible natural resources, as Chile considered that some EU fishing vessels were making excessive catches of swordfish beyond the Chilean exclusive economic zone and that, because swordfish are a highly migratory species, the EU fishing vessels were harming Chilean fisheries and the preservation of the species. Chile also responded to the EU by lodging a complaint before the International Tribunal for the Law of the Sea (ITLOS) in Hamburg, alleging that the EU was not taking appropriate measures to control the activity of EU fishing vessels in certain international waters, in breach of the rules of the United Nations Convention on the Law of the Sea (UNCLOS) that relate to the conservation and sustainable exploitation of fishing resources (ITLOS 2000). This dispute, which was brought before two different international bodies from different perspectives, was of great legal interest, showing a practical example of the so-called fragmentation of the international legal system (Orellana 2002). In any case, the dispute was resolved by a bilateral agreement between the parties: Chile agreed to repeal the previous ban that it had imposed on the landing of swordfish by EU fishing vessels, while also allowing a reasonable maximum annual landing quota (WTO 2001c).

Another interesting dispute over the sustainable exploitation of fisheries is the *EU – Herring* case. The request for consultations was submitted on 4 November 2013 by Denmark in respect of the Faroe Islands, which is a self-governing territory under Danish sovereignty and not part of the EU (WTO 2013). The Faroe Islands had decided to unilaterally extend their herring quota, ignoring the allocations set for herring fishing under a previous agreement between the North Atlantic coastal states and claiming that the presence of herring in waters close to their coasts had increased greatly, probably owing to the displacement of herring towards more northern waters as a result of global warming (Fernández Egea 2014, p. 3). The EU reacted by applying its rules for the imposition of restrictions on countries that allow unsustainable fishing. In particular, the EU prohibited imports of herring caught by vessels from the Faroe Islands (or under their control) and the landing of such vessels in EU ports. The complainant argued that such trade restrictions imposed by the EU were incompatible with basic principles of GATT 1994. The EU sought to justify its measures in light of Art. XX(g) of GATT 1994. The panel was established on 26 February 2014. Ultimately, however, the parties reached a mutually agreed solution that was notified to the WTO in August 2014 (WTO 2014). The negotiated solution set the herring quota for the Faroe Islands at 40,000 tonnes, which was higher than the 31,000 tonnes initially allocated under the previous agreement, but much lower than the Faroe Islands' initial aim. Without prejudice to the negotiated solution, some scholars conclude that the EU measures could be clearly justified under Art. XX(g) of GATT 1994 and its *chapeau*, noting that such EU measures were designed and applied taking WTO jurisprudence into account, particularly the *US—Shrimp* case (Fernández Egea 2014; Auld 2018).

Another recent case, *EU—Palm oil (Indonesia)*, is also of particular interest for this chapter. The complaint was filed by Indonesia on 9 December 2019 (WTO 2019). The EU had determined that biofuel obtained from palm oil should not generally be considered a sustainable biofuel, since it observes that most palm plantations in Indonesia (and other tropical countries) are expanding at the expense of the deforestation of primary forests, which are important carbon sinks. Thus, the EU counts the emissions attributable to indirect land-use change (ILUC) from deforestation in the carbon footprint of biofuel from palm oil. Indonesia claims that such EU measures, adopted unilaterally and relating to non-product-related PPMs, are incompatible with basic principles of GATT 1994 and other WTO rules and considers that the EU is unduly penalizing the use by developing countries of natural resources located in their own territories (WTO 2019, paras. 29–34). The panel was established on 29 July 2020 and has before it a complex case that mixes trade, climate change, development and sustainable use of biodiversity. While the case certainly focuses on the calculation of the carbon footprint, the increasing destruction of primary tropical forests for the expansion of certain crops (such as palm or sugar cane) obviously has a host of other devastating effects on the preservation of biodiversity. The possible legal solution of the case is highly controversial among legal scholars (Limenta 2020; Mitchell and Merriman 2020).

Pending the resolution of this complex case, it must be recognized that the WTO dispute settlement system has been contributing to improved conciliation between trade and the conservation and sustainable use of biodiversity. However, there are limitations to this interpretative approach. It has been formulated in relation to the specific situations raised, which have, until now, mainly focused on the conservation and sustainable use of marine animals in fishing activities. As some scholars observe, there are still many legal ambiguities or uncertainties, particularly regarding the jurisdictional scope of various paragraphs of Art. XX of GATT 1994, the conditions of its *chapeau* and the specific requirements to justify certain unilateral trade restrictions in response to climate change and to protect biodiversity abroad (Condon 2009; Trachtman 2017, p. 309; Calle Saldarriaga 2018, p. 252). Thus, current WTO jurisprudence does not remove the need for more ambitious trade negotiations to better connect trade and sustainability.

5.4 Multilateral and Plurilateral Negotiations in the WTO

The multilateral trade negotiations launched by the WTO with the Doha Round include several aspects on trade and the environment. Held in Doha (Qatar) in 2001, the Fourth WTO Ministerial Conference, which initiated the round, adopted a Ministerial Declaration that provided, mainly based on previous work and discussion in the CTE, for negotiations on, for example, the relationship between existing WTO rules and specific trade obligations (STOs) set out in MEAs; procedures for regular information exchange between MEA Secretariats and the relevant WTO committees; the reduction or, as appropriate, elimination of tariff and non-tariff barriers to environmental goods and services; and the clarification and improvement of WTO disciplines on fisheries subsidies (WTO 2001d, paras. 31–33). As is well known, however, the Doha Round has so far produced few results, because it is currently very difficult to reach large-scale consensus on global trade rules among advanced, emerging and developing countries.

Currently, the main negotiating efforts of the Doha Round on environmental issues focus on fisheries subsidies. As we have seen, this issue was included in the Doha Ministerial Declaration where it was linked, at least in part, to the conservation and sustainable use of biodiversity, since various international institutions and the CTE had warned that fisheries subsidies often contribute to the overexploitation of fish resources. The environmental objective of the negotiations was detailed at the Sixth WTO Ministerial Conference, held in Hong Kong (China) in 2005, which also specified the need to integrate an "appropriate and effective special and differential treatment for developing and least-developed members", taking into account "the importance of this sector to development priorities, poverty reduction, and livelihood and food security concerns" (WTO 2005, Annex D, para. 9).

The negotiations, which have been very difficult in practice because fishing is a particularly sensitive economic sector for many countries, received a renewed boost in 2015, thanks to the SDGs approved by the General Assembly of the United Nations (UN). SDG 14, namely to "[c]onserve and sustainably use the oceans, seas and marine resources for sustainable development", includes the following Target 14.6 to be achieved "taking into account ongoing WTO negotiations" on fisheries subsidies:

> By 2020, prohibit certain forms of fisheries subsidies which contribute to overcapacity and overfishing, eliminate subsidies that contribute to illegal, unreported and unregulated fishing [IUU fishing] and refrain from introducing new such subsidies [...].

This explicit connection between the SDGs and the WTO negotiations shows another way to overcome the traditional fragmentation between various sectors of the international legal system. The European Commission, in a 2018 concept paper on WTO reform, highlights this example of synergy between the UN and the multilateral trading system and argues that it should be taken as a model for further possible contributions of the WTO to "the sustainability objectives of the global community" (European Commission 2018a, p. 6, section I.C).

The Eleventh WTO Ministerial Conference, which took place in Buenos Aires (Argentina) in 2017, approved a brief Ministerial Decision on fisheries subsidies, in which the members committed to continue the negotiations with a view to adopting an agreement by the next Ministerial Conference (WTO 2017a, para. 1).

The Twelfth WTO Ministerial Conference was set to take place in Nur-Sultan (Kazakhstan) from 8 to 11 June 2020. However, the COVID-19 pandemic led to its cancellation. On 29 May 2020, Kazakhstan offered to host the conference in June 2021, but no final decision has yet been taken (WTO 2020a).

In any case, the current chair of WTO negotiations on fisheries subsidies, Ambassador Santiago Wills of Colombia, is trying to conclude the negotiations soon and, on 25 June 2020, released a draft consolidated text of the agreement (WTO 2020b). At the 21 July 2020 meeting of the Negotiating Group on Rules, heads of the WTO delegations set a work programme to reach an agreement by the end of 2020 (WTO 2020c). The draft still contains many provisions with various options in brackets and there are important issues on which a consensus has yet to be reached (Tipping and Irschlinger 2020).

Apart from the difficult multilateral negotiations noted above, there have been some successful plurilateral negotiations between a limited number of WTO members that take environmental aspects into account. Thus, the plurilateral Government Procurement Agreement (GPA) of 1994 was updated in 2012, thanks to the leadership of the EU and the US (under the presidency of Barack Obama). This is the only major agreement reached in Marrakesh that has been revised in depth, probably because it is an agreement concluded almost exclusively between advanced economies. GPA 2012, which entered into force on 6 April 2014, has corrected the absence of environmental references in the original GPA 1994 (Anderson and Muller 2017, p. 967).

Specifically, GPA 2012 contains a new provision, Art. X:6, which provides for greater certainty over the possibility of using “technical specifications to promote the conservation of natural resources or protect the environment”. Furthermore, the term “technical specifications” is defined, in Art. I:u), as a tendering requirement that “lays down the characteristics of goods or services to be procured, including quality, performance, safety and dimensions, or the processes and methods for their production or provision”, confirming that the latter can encompass both product-related PPMs and non-product-related PPMs. In addition, under Art. XXII:8.a)iii, the Committee on Government Procurement adopted a work programme on “the treatment of sustainable procurement” and the WTO organized a symposium on sustainable public procurement in Geneva on 22 February 2017 (WTO 2017b). The revised GPA 2012 is therefore a good example for the possible updating of other WTO rules.

5.5 Provisions in Regional Trade Agreements

Art. XXIV GATT 1994 (on "Customs Unions and Free-trade Areas") is the main legal basis for the coexistence between the multilateral trading system and other regional trade agreements. The multilateral and plurilateral agreements of the WTO have always coexisted with regional trade agreements. However, the limited results achieved in the multilateral trade negotiations of the Doha Round appear to have led, in recent years, to a major proliferation of regional and mega-regional trade agreements, which have been concluded in particular by the largest economies with the most like-minded countries. These recent regional trade agreements (RTAs) are characterized by their broad content (goods, services, intellectual property, investment, government procurement, etc.), and they usually contain, particularly in those led by advanced economies, numerous provisions on the environment and the conservation and sustainable use of biodiversity, which represent a paradigm shift in the regulation of international trade (Anuradha 2017, p. 241).

A good example is the Comprehensive and Progressive Agreement for Trans-Pacific Partnership (CPTPP). The origin of this agreement lies in the Trans-Pacific Partnership (TPP), which was promoted by the US during the Obama presidency and involved eleven other Pacific Rim countries (with the very significant absence of China). The TPP was signed in Auckland (New Zealand) on 4 February 2016. However, on 23 January 2017, Donald Trump decided that the US would not ratify the TPP and the agreement did not come into force (Ji and Rana 2019). The rest of the TPP negotiating countries, however, decided to go ahead with the initiative and they introduced certain changes. The formal signing ceremony of the CPTPP was held on 8 March 2018 in Santiago (Chile) and it entered into force on 30 December 2018. Currently, the CPTPP has seven States Parties: Australia, Canada, Japan, Mexico, New Zealand, Singapore and Vietnam (Australian Government 2020).

In the preamble to the CPTPP, the parties recognize "their inherent right to regulate" and protect "legitimate public welfare objectives, such as public health, safety, the environment, [and] the conservation of living or non-living exhaustible natural resources [...]". The CPTPP includes numerous environmental provisions, particularly in Chapter 20 on "environment", which comprises, in addition to general commitments, specific provisions on, for instance, "multilateral environmental agreements" (Art. 20.4); "trade and biodiversity" (Art. 20.13), which starts recognizing "the importance of conservation and sustainable use of biological diversity and their key role in achieving sustainable development"; "invasive alien species" (Art. 20.14); "transition to a low emissions and resilient economy" (Art. 20.15); "marine capture fisheries" (Art. 20.16), which provides for concrete measures for the protection of sharks, marine turtles, seabirds and marine mammals and includes a specific regulation on fisheries subsidies, which has anticipated the aforementioned negotiations carried out under the WTO and underscores the ability of RTAs to tackle issues that have stalled within the multilateral trading system; and "conservation and trade" (Art. 20.17), which recognizes that illegal trade in "wild fauna and flora" undermines efforts "to conserve and sustainably manage those natural resources". The CPTPP provides that its general mechanism for dispute settlement, which includes the establishment of a panel of experts, also applies to possible disputes on the aforementioned chapter on "environment", providing that, in such cases, "panellists other than the chair shall have expertise or experience in environmental law or practice" (Art. 28.9.5(b)).

The CPTPP can therefore be regarded as one of the most ambitious environmental RTAs (Chen 2019, p. 22) and its regulatory model could inspire other international trade agreements and future reforms of the multilateral trading system. Some authors observe, however, that only practice can demonstrate the true effectiveness of this type of environmental provisions, whose purpose could be to provide a patina of environmental concern to treaties aimed primarily at expanding trade and investment (Meidinger 2019, p. 195).

Although the Transatlantic Trade and Investment Partnership (TTIP) negotiated between the US and the EU has been put on hold under the Trump presidency, the EU has maintained an ambitious agenda of trade negotiations with third countries and typically designates its most recent agreements as a "new generation of free trade agreements" (European Commission 2019a, p. 13). Such negotiations include, for example, those with South Korea; Colombia–Peru–Ecuador; Central America; Canada, with the agreement known as the Comprehensive Economic and Trade Agreement (CETA); Singapore; Japan; Vietnam; and Mercosur (European Commission 2020a). All these negotiations are characterized by their broad scope, and the concluded agreements usually include a chapter dedicated to "trade and sustainable development" (TSD), with various social and environmental provisions. The European Green Deal Communication insists that all RTAs concluded by the EU must incorporate a TSD chapter (European Commission 2019b).

A relevant example is the TSD chapter included in the EU–Mercosur Agreement in principle announced on 28 June 2019 (European Commission 2019c), particularly considering the presence of Brazil, which is one of the world's richest biological megadiversity countries. The TSD chapter in question includes 18 articles, with substantive provisions on, *inter alia*: "multilateral environmental agreements" (Art. 5); "trade and climate change" (Art. 6), which contains explicit references to the Paris Agreement; "trade and biodiversity" (Art. 7), which recognizes the importance of the conservation and sustainable use of biological diversity consistent with the CBD, CITES and the International Treaty on Plant Genetic Resources for Food and Agriculture (and the decisions adopted thereunder) and which also encourages "trade in natural resource-based products obtained through a sustainable use of biological resources or which contribute to the conservation of biodiversity"; "trade and sustainable management of forests" (Art. 8), providing that each party shall encourage "trade in products from sustainably managed forests"; and "trade and sustainable management of fisheries and aquaculture" (Art. 9), specifying, for example, that each party shall "act consistent with the principles of the FAO Code of Conduct for Responsible Fisheries", thereby contributing to the strengthening of this international instrument. From an institutional point of view (Arts. 14–18), the TSD chapter, as is usual in other RTAs of the EU with third countries, provides for the creation of a TSD subcommittee and a specific mechanism for dispute resolution, without sanctions, and softer than other mechanisms established for other chapters. This is an important difference with respect to the CPTPP, which, as indicated above, has a general mechanism for dispute resolution that is also applicable, as a last resort, to the chapter on "environment".

The systematic inclusion of a TSD chapter, with numerous provisions on biodiversity, in the RTAs concluded by the EU is laudable. However, there is growing concern over the true relevance of such provisions in practice (Laurans 2020). In February 2018, the European Commission identified "15 action points" to make TSD chapters more effective (European Commission 2018b, p. 2).

A report on methodologies and indicators prepared for the European Commission in 2018 to assess the impact of trade liberalization on biodiversity points out the need to complement traditional qualitative monitoring with quantitative methodologies and "a more systematic use of biodiversity indicators" (Kuik et al. 2018, p. 8).

The Communication on the EU Biodiversity Strategy for 2030, presented on 20 May 2020, states that the European Commission "will ensure full implementation and enforcement of the biodiversity provisions in all trade agreements", including through the new position of EU Chief Trade Enforcement Officer (CTEO) (European Commission 2020b, p. 21, para. 4.2.2).

An expert report on trade and biodiversity presented to the European Parliament in June 2020 notes that the inclusion of numerous provisions on biodiversity in TSD chapters currently has "uncertain" impacts, since such provisions are hardly effective and the challenge "is to make them more effective and enforceable" (Bellora et al. 2020, pp. 19 and 32). The report also suggests less "legal inflation" of RTAs with generic provisions on biodiversity as well as the inclusion in TSD chapters of more enforceable clauses, which should be quantifiable and guaranteed through transparent and automatic "sanctions" in case of non-compliance (Bellora et al. 2020, pp. 6, 20 and 32). In any case, the EU should also increase its support to developing countries in the implementation of such TSD chapters.

5.6 Conclusion

This chapter has sought to show how consideration of the conservation and sustainable use of biodiversity in the rules of the multilateral trading system and other international agreements on trade in goods has evolved. This evolution can be summarized very succinctly as a long, complex and incomplete transition from fragmentation to systemic interconnection.

The WTO dispute settlement system has contributed significantly to the search for legal conciliation between trade and the environment, dealing with some cases that specifically concern trade measures for the conservation and sustainable use of biodiversity, such as the landmark *US—Shrimp* case. The WTO's adjudicative bodies have promoted an evolutionary interpretation of WTO rules, giving consideration to other international instruments on the environment and biodiversity. In any case, there are still legal uncertainties and discussions about the proper interpretation of certain provisions, concerning, in particular, the jurisdictional scope of various paragraphs of Art. XX of GATT 1994 and the conditions of its *chapeau*.

This jurisprudential approach should have been accompanied by an appropriate updating of WTO rules through multilateral negotiations. Unfortunately, however, the required consensus and the growing tensions among advanced, emerging and developing countries have so far prevented any major new global trade rules. Most of the issues to be negotiated on trade and the environment proposed by the CTE have been blocked. The efforts of the Doha Round currently focus on the adoption of a new multilateral agreement regarding the most harmful fisheries subsidies and is explicitly linked to Target 14.6 of the SDGs. The plurilateral negotiations to revise the GPA 1994 by the new GPA 2012, which has incorporated provisions for sustainable public procurement, can be considered a successful updating of WTO rules, but it is an agreement concluded mainly by advanced economies and it is difficult to replicate in multilateral agreements that bind, as a single undertaking, all WTO members.

In contrast to the notable immobility of the WTO agreements, there is a new approach to relations between trade and the environment in recent RTAs, which include numerous detailed provisions on sustainable development, MEAs, biodiversity, sustainable management of forests and fisheries, etc. This trend in RTAs is encouraging. It clearly surpasses the traditional focus of the multilateral trading system on trade issues and promotes a new, more holistic regulatory model, making explicit numerous systemic interconnections between trade rules and environmental rules, which can generate positive synergies between them. Ideally, a future in-depth updating of the WTO could draw inspiration from this model.

The greening of international trade agreements is good, but it is no time for complacency. The main objectives of RTAs remain the promotion of trade and investment between the parties. It is necessary to ensure that the proliferation of environmental provisions is not merely an ornament but has a real impact in practice by improving and strengthening their monitoring (including a systematic use of biodiversity indicators) and implementation mechanisms.

It is worth remembering that the effective application of environmental regulations, which often go against the most immediate economic interests, tends to be particularly problematic, even within advanced legal systems such as the EU (Krämer 2016). Perhaps the current crisis caused by Covid-19 will contribute in some way towards a greater appreciation of the conservation and sustainable use of biodiversity, redressing the balance between, on the one hand, the search for efficiency in economic terms and, on the other, the resilience of humanity and the respect for nature.

References

Anderson, R. D., & Muller, A. C. (2017). The revised WTO Agreement on Government Procurement (GPA): Key design features and significance for global trade and development. *Georgetown Journal of International Law, 48*, 949–1008.

Ankersmit, L. J., & Lawrence, J. C. (2012). The future of environmental labelling: US – Tuna II and the scope of the TBT. *Legal Issues of Economic Integration, 39*(1), 127–147.

Anuradha, R. V. (2017). WTO to TPP: Evolution of environmental provisions in trade agreements. In J. Chaisse et al. (Eds.), *Paradigm shift in international economic law rule-making: TPP as new model for trade agreements?* (pp. 241–245). Singapore: Springer.

Appleton, A. E. (2015). Product labelling: What has the appellate body wrought? *ICTSD – Biores, 9* (8). Retrieved Sept 27, 2020, from https://ictsd.iisd.org/bridges-news/biores/news/product-labelling-what-has-the-appellate-body-wrought

Auld, K. (2018). Trade measures to prevent illegal fishing and the WTO: An analysis of the settled Faroe Islands dispute. *World Trade Review, 17*(4), 665–692.

Australian Government. (2020). About the Comprehensive and Progressive Agreement for Trans-Pacific Partnership (CPTPP). Retrieved Sept 27, 2020, from https://www.dfat.gov.au/trade/agreements/in-force/cptpp/Pages/comprehensive-and-progressive-agreement-for-trans-pacific-partnership

Baroncini, E., & Brunel, C. (2020). A WTO safe harbour for the dolphins: The second compliance proceedings in the US – Tuna II (Mexico) case. *World Trade Review, 19*(2), 196–215.

Bellora, C., Bureau, J. C., Bayramoglu, B., Gozlan, E., & Jean, S. (2020). Trade and biodiversity. Research report requested by the European Parliament's Committee on International Trade (INTA). Retrieved Sept 27, 2020, from https://www.europarl.europa.eu/RegData/etudes/IDAN/2020/603494/EXPO_IDA(2020)603494_EN.pdf

Calle Saldarriaga, M. A. (2018). Sustainable production and trade discrimination: An analysis of the WTO jurisprudence. *Anuario Colombiano de Derecho Internacional, 11*, 221–258.

Charnovitz, S. (2002). The law of environmental "PPMs" in the WTO: Debunking the myth of illegality. *Yale Journal of International Law, 27*(1), 59–110.

Chen, I. J. (2019). An analysis of establishing regional environmental governance through a mega regional trade agreement: The Asia–Pacific practice. *Vermont Journal of Environmental Law, 20*, 1–31.

Condon, B. (2009). Climate change and unresolved issues in WTO law. *Journal of International Economic Law, 12*(4), 895–926.

Conrad, C. R. (2011). *Processes and production methods (PPMs) in WTO law: Interfacing trade and social goals*. Cambridge: Cambridge University Press.

Cooreman, B. (2017). *Global environmental protection through trade: A systematic approach to extraterritoriality*. Cheltenham: Edward Elgar.

Costanza, R., Audley, J., Borden, R., Ekins, P., Folke, C., Funtowicz, S. O., et al. (1995). Sustainable trade: A new paradigm for world welfare. *Environment: Science and Policy for Sustainable Development, 37*(5), 16–44.

Deke, O. (2008). *Environmental policy instruments for conserving global biodiversity*. Berlin: Springer.

Esty, D. C. (1993). *Greening the GATT: Trade, environment and the future*. Oxford: Oxford University Press.

European Commission. (2018a). EU concept paper on WTO reform. Retrieved Sept 27, 2020, from https://trade.ec.europa.eu/doclib/docs/2018/september/tradoc_157331.pdf

European Commission. (2018b). Feedback and way forward on improving the implementation and enforcement of Trade and Sustainable Development chapters in EU Free Trade Agreements. Non-paper of the Commission Services. Retrieved Sept 27, 2020, from http://trade.ec.europa.eu/doclib/docs/2018/february/tradoc_156618.pdf

European Commission. (2019a). 2019 Report on implementation of EU free trade agreements 1 January 2018 – 31 December 2018. Retrieved Sept 27, 2020, from https://trade.ec.europa.eu/doclib/docs/2019/october/tradoc_158387.pdf

European Commission. (2019b). The European Green Deal, COM(2019) 640 final, 11 December 2019.

European Commission. (2019c). EU–Mercosur trade agreement: The agreement in principle and its texts. Brussels, 12 July 2019. Retrieved Sept 27, 2020, from https://trade.ec.europa.eu/doclib/press/index.cfm?id=2048

European Commission. (2020a). Overview of FTA and other trade negotiations. Updated August 2020. Retrieved Sept 27, 2020, from https://trade.ec.europa.eu/doclib/docs/2006/december/tradoc_118238.pdf

European Commission. (2020b). EU Biodiversity Strategy for 2030, COM(2020) 380 final, 20 May 2020.

Fernández Egea, R. M. (2008). *Comercio de mercancías y protección del medio ambiente en la OMC*. Madrid: Marcial Pons.

Fernández Egea, R. M. (2014). Cambio climático y la sostenibilidad de los recursos pesqueros del Mar del Norte: La controversia comercial entre la Unión Europea y las Islas Feroe. *Revista del Instituto Español de Estudios Estratégicos, 4*, 1–22.

Fernández Pons, X., & Baroncini, E. (2004). The WTO context for sustainability labelling and certification. In M. Campins Eritja (Ed.), *Sustainability labelling and certification* (pp. 125–167). Madrid: Marcial Pons.

GATT. (1970). Border tax adjustments. Report of the Working Party adopted on 2 December 1970, L/3464.

GATT. (1991). US – Restrictions on imports of tuna (Mexico) ("US – Tuna (Mexico)"). Panel Report, DS21/R.

GATT. (1994). US – Restrictions on imports of tuna (EEC) ("US – Tuna (EEC)"). Panel Report, DS29/R.

Holzer, K. (2017). II.13 Processes and production methods (PPMs). In T. Cottier & K. Nadakavukaren Schefer (Eds.), *Elgar encyclopedia of international economic law* (pp. 189–191). Cheltenham: Edward Elgar.

Howse, R. (2002). The appellate body rulings in the shrimp/turtle case: A new legal baseline for the trade and environment debate. *Columbia Journal of Environmental Law, 27*(2), 491–591.

Hurlock, M. H. (1992). The GATT, US law and the environment: A proposal to amend the GATT in light of the Tuna/Dolphin decision. *Columbia Law Review, 92*(8), 2098–2161.

ITLOS. (2000). Case concerning the conservation and sustainable exploitation of swordfish stocks in the South-Eastern Pacific Ocean (Chile/EU), Order 2000/3 of 20 December 2000. *ITLOS Reports, 2000*, 148.

Ji, X., & Rana, P. B. (2019). A deal that does not die: The United States and the rise, fall and future of the (CP)TPP. *Pacific Focus, 34*(2), 230–255.

Juste Ruiz, J. (2003). Protección del medio ambiente y comercio internacional. *Cursos Euromediterráneos Bancaja de Derecho Internacional, 7*, 341–456.

Krämer, L. (Ed.). (2016). *Enforcement of environmental law*. Cheltenham: Edward Elgar.

Kuik, O., Kettunen, M., van Vliet, J., Colsa, A., & Illes, A. (2018). Trade liberalisation and biodiversity: Scoping study on methodologies and indicators to assess the impact of trade liberalisation on biodiversity (ecosystems and ecosystem services). Final report for the European Commission (DG ENV) (ENV.F.1/FRA/2014/0063). Retrieved Sept 27, 2020, from https://ec.europa.eu/environment/nature/pdf/scoping_study.pdf

Laurans, Y. (2020). The EU–Mercosur trade agreement: Can a trade treaty be greened? Institut du développement durable et des relations internationales (IDDRI). Retrieved Sept 27, 2020, from https://www.iddri.org/en/publications-and-events/blog-post/eu-mercosur-trade-agreement-can-trade-treaty-be-greened

Limenta, M. (2020). Palm oil for fuels: WTO rules and environmental protection. *Global Trade and Customs Journal, 15*(7), 331–339.

Maggio, A. R. (2017). *Environmental policy, non-product related process and production methods and the law of the World Trade Organization*. Cham: Springer.

McDorman, T. L. (1992). The 1991 US–Mexico GATT Panel report on Tuna and Dolphin: Implications for trade and environment conflicts. *North Carolina Journal of International Law and Commercial Regulation, 17*(3), 461–488.

Meidinger, E. (2019). The Trans-pacific partnership agreement and environmental regulation. In B. Kingsbury et al. (Eds.), *Megaregulation contested: Global economic ordering after TPP* (pp. 175–195). Oxford: Oxford University Press.

Mitchell, A. D., & Merriman, D. (2020). Indonesia's WTO challenge to the European Union's renewable energy directive: Palm oil and indirect land-use change. *Trade, Law and Development, 12*(2), 1–56.

Nordström, H., & Vaughan, S. (1999). *Trade and the environment*. Geneva: WTO.

OECD. (1994). *Trade and environment: Processes and production methods*. Paris: OECD.

Orellana, M. A. (2002). The Swordfish dispute between the EU and Chile at the ITLOS and the WTO. *Nordic Journal of International Law, 71*(1), 55–81.

Sifonios, D. (2018). *Environmental process and production methods (PPMs) in WTO law*. Cham: Springer.

Sinha, M. (2013). An evaluation of the WTO Committee on Trade and Environment. *Journal of World Trade, 4*(6), 1285–1322.

Tipping, A., & Irschlinger, T. (2020). WTO negotiations on fisheries subsidies: What's the state of play? International Institute for Sustainable Development (IISD). Retrieved Sept 27, 2020, from https://www.iisd.org/system/files/2020-08/wto-negotiations-fisheries-state-play_0.pdf

Trachtman, J. P. (2017). WTO trade and environment jurisprudence: Avoiding environmental catastrophe. *Harvard International Law Journal, 58*(2), 273–309.

WTO. (1996). US – Import prohibition of certain shrimp and shrimp products ("US – Shrimp"). Request for consultations by India, Malaysia, Pakistan and Thailand, WT/DS58/1.

WTO. (1998). US – Import prohibition of certain shrimp and shrimp products ("US – Shrimp"). Appellate Body Report, WT/DS58/AB/R.

WTO. (2000a). US – Import prohibition of certain shrimp and shrimp products ("US – Shrimp"). Recourse to Art. 21.5 of DSU by Malaysia, WT/DS58/17.

WTO. (2000b). Chile – Measures affecting the transit and importing of swordfish ("Chile – Swordfish"). Request for consultations by the EC, WT/193/1.

WTO. (2001a). US – Import prohibition of certain shrimp and shrimp products ("US – Shrimp"). Recourse to Art. 21.5 of DSU by Malaysia. Panel Report, WT/DS58/RW.

WTO. (2001b). US – Import prohibition of certain shrimp and shrimp products ("US – Shrimp"). Recourse to Art. 21.5 of DSU by Malaysia. Appellate Body Report, WT/DS58/AB/RW.

WTO. (2001c). Chile – Measures affecting the transit and importing of swordfish ("Chile – Swordfish"). Arrangement between the EC and Chile, WT/DS193/3.

WTO. (2001d). Ministerial Declaration adopted in Doha on 14 November 2001, WT/MIN(01)/DEC.

WTO. (2005). Ministerial Declaration adopted in Hong Kong on 18 December 2005, WT/MIN(05)/DEC.

WTO. (2008). US – Measures concerning the importation, marketing and sale of tuna and tuna products ("US – Tuna II (Mexico)"). Request for consultations by Mexico, WT/DS381/1.

WTO. (2011). US – Measures concerning the importation, marketing and sale of tuna and tuna products ("US – Tuna II (Mexico)"). Panel Report, WT/DS381/R.

WTO. (2012). US – Measures concerning the importation, marketing and sale of tuna and tuna products ("US – Tuna II (Mexico)"). Appellate Body Report, WT/DS381/AB/R.

WTO. (2013). EU – Measures on Atlanto-Scandian herring ("EU – Herring"). Request for consultations by Denmark in respect of the Faroe Islands, WT/DS469/1.

WTO. (2014). EU – Measures on Atlanto-Scandian herring ("EU – Herring"). Joint communication by Denmark in respect of the Faroe Islands and the EU, WT/DS469/3.

WTO. (2017a). Fisheries subsidies. Ministerial Decision adopted in Buenos Aires on 13 December 2017, WT/MIN(17)/64.

WTO. (2017b). Symposium on sustainable procurement. Geneva, 22 February 2017. Retrieved Sept 27, 2020, from https://www.wto.org/english/tratop_e/gproc_e/gp_symp_22feb17_e.htm

WTO. (2018). US – Measures concerning the importation, marketing and sale of tuna and tuna products ("US – Tuna II (Mexico)"). Recourse to Art. 21.5 of DSU by the US. Appellate Body Report, WT/DS381/AB/RW/USA.

WTO. (2019). EU – Certain measures concerning palm oil and oil palm crop-based biofuels ("EU – Palm oil (Indonesia)"). Request for consultations by Indonesia, WT/DS593/1.

WTO. (2020a). WTO members discuss Kazakhstan's offer to host 12th Ministerial Conference in June 2021. Retrieved Sept 27, 2020, from https://www.wto.org/english/news_e/news20_e/gc_29may20_e.htm

WTO. (2020b). Fisheries subsidies. Draft consolidated text, RD/TN/RL/126.

WTO. (2020c). WTO members set work program for text-based phase of fisheries subsidies talks. Retrieved Sept 27, 2020, from https://www.wto.org/english/news_e/news20_e/fish_21jul20_e.htm

Yavitz, L. (2001). The WTO and the environment: The Shrimp case that created a new world order. *Journal of Natural Resources and Environmental Law, 16*(3), 203–255.

Young, M. A. (2014). Trade measures to address environmental concerns in faraway places: Jurisdictional issues. *Review of European, Comparative and International Environmental Law, 23*(3), 302–317.

Chapter 6
Is Everything Marketable in International Trade? Public Health Issues in International Trade of Wildlife

Josep Maria de Dios Marcer

Contents

6.1 Introduction 102
6.2 Protected Species in the Context of International Trade: Are They Products, Goods, or Something Different? 103
6.2.1 Wild Species and International Trade. The Regulations of the World Trade Organization 104
6.2.2 Protection of Endangered Species of Wild Fauna and Flora and Trade Within Regional Organizations. The European Union 106
6.3 Public Health and the International Trade in Endangered Species 108
6.3.1 The Concept and the Scope of Public Health 109
6.3.2 The Relationship Between Public Health and International Trade 110
6.4 CITES: An Instrument for the Protection of Public Health? 112
6.5 Conclusion 114
References 114

Abstract The scientific community has noted on a regular basis the link between the international trade of protected species in the natural environment and public health. An example is the first coronavirus (SARS-CoV-1) pandemic, which is believed to have begun in wildlife markets and to have expanded by means of the international wildlife trade. The goal of this chapter is to apply a legal perspective to the relationship between three elements: the limits on the international trade of protected species, the relationship between international trade and the international protection of these species, and the relationship between the international trade of species and public health. To this end, the chapter analyses different international conventions and the action of international organizations. The key issue is to determine whether the CITES convention is the most appropriate place to include norms for the protection of public health from the potential dangers of the international trade in protected species, or whether these norms should be maintained inside the regulations governing international trade, bearing in mind that the remit of

J. M. de Dios Marcer (✉)
Universitat Autònoma de Barcelona, Faculty of Law, Bellaterra, Spain
e-mail: josepmaria.dedios@uab.cat

M. Campins Eritja, T. Fajardo del Castillo (eds.), *Biological Diversity and International Law*, https://doi.org/10.1007/978-3-030-72961-5_6

CITES does not extend beyond the protection of endangered species by means of restricting their international commercialization.

6.1 Introduction

Scientists have attributed the origin of the COVID-19 pandemic to wild animals. The Convention on International Trade in Endangered Species of Wild Fauna and Flora (CITES) (1973) introduces limits on the trading of wildlife. The pangolin, for example, strongly implicated in the COVID-19 pandemic, has enjoyed the maximum degree of protection since 2017; however, it continues to be bought and sold in certain Asian markets (not always legally) due to its special medicinal or gastronomic value.

The case of the pangolin highlights three circumstances: first, that restrictions on trade on biological grounds do not stop the distribution of protected species; second, that there is a direct link between wildlife and public health; and third, that both wildlife and public health are fragile. Zoonotic diseases are not an alien element in today's world, and they manifest themselves in many different ways: via mosquitoes in the case of the Nile Virus, via bats in the case of the 2002 SARS-CoV-1, or allegedly via small land mammals in the case of COVID-19. Indeed, in the wake of the COVID-19 pandemic, the protection of wildlife species, the trading of these species and the impact of this trade on public health are among the pressing issues facing the global community today.

The purpose of this study is to present a legal analysis of these concerns and to assess what means are available, or need to be introduced, to ensure that protection and trading can coexist in a harmonious way. From the perspective of protection, the CITES treaty aims to safeguard wild fauna and flora species in danger of extinction by limiting their commercialization. From the perspective of international trade, these species (or products made from them) are considered as marketable goods, and so the rules governing international trade are applicable. However, trade in species is a highly sensitive matter because of the nature of the item traded and because of the spread of diseases that it may cause. Thus, both the species themselves and the public health of the environment into which they are introduced are in need of protection.

This chapter will explore the legal bases of the wildlife trade, the different forms of protection, and the interaction between the two.

6.2 Protected Species in the Context of International Trade: Are They Products, Goods, or Something Different?

The question to be determined in this section is whether protected wildlife species are objects, products or goods that may be traded, or whether they have a special status that should render their commerce illegal. The study will focus on two areas that have a particularly strong impact on international trade: the WTO-GATT, and the EU.

The CITES Convention is an instrument that establishes measures to protect wild fauna and flora, and introduces limits and conditions for international trade in species whose *excessive exploitation* may represent a threat to their survival (CITES 1973, preamble). However, CITES does not regulate the trade of protected species or prohibit it; rather, it subjects this commerce to certain conditions (Scanlon 2020a), based on the incorporation of mandatory minimum standards (CITES 1973, art. XIV). The restrictiveness of the measures regarding trade depend on the species in question and its inclusion in Appendix I, II or III of the Convention, Appendix I representing the species under greatest threat (CITES 1973, art.II). The Convention also allows States the option of introducing more severe restrictions.

Therefore, one could say that CITES is a convention with biological objectives, which focuses on the preservation of threatened species by introducing minimum conditions and limits on their trade.[1] Therefore, its aim is simply to combine international trade with the protection of wild fauna and flora. For other issues, domestic or international rules will apply (CITES 1973, art.XIV.2).

The intended protection is to be achieved via cooperation at national and international levels. National authorities adopt measures for providing this protection, and at the same time, States have established a system of international cooperation to make it effective.

The conclusion is that the text of the Convention makes international trade in protected species conditional on the fulfilment of certain requirements. To use legal terminology, these are imperative *de minimis* rules (CITES 1973, art. XIV), of an international origin, which are necessarily applicable whatever legislation a particular State may apply to the purchase or sale of any of these species. The cooperation of the authorities facilitates its implementation when trade is between Member States, but if one of the States concerned is not a party to CITES, Article X of the Convention will apply.[2]

---

[1]In the Resolution adopted by the General Assembly on 27 July 2012 (66/288 The Future We Want) the UN considers CITES as an international agreement that stands at the intersection between trade, the environment and development, promoting the conservation and sustainable use of biodiversity (see point n° 203 of the Resolution).

[2]Where export or re-export is to, or import is from, a State not a Party to the present Convention, comparable documentation issued by the competent authorities in that State which substantially complies with the requirements of the present Convention for permits and certificates may be accepted in lieu thereof by any Party.

6.2.1 *Wild Species and International Trade. The Regulations of the World Trade Organization*

From the perspective of the WTO, the concept of "product" refers to goods traded and subject to GATT 1994 and other multilateral agreements on trade in goods. This concept also covers wildlife species, including those under special protection. Therefore, the WTO regulations apply to these species, which are considered tradable. The WTO incorporates two types of rules: one designed to promote sustainability, and the other to protect public health.

Measures for preserving sustainability have already been studied in Chap. 5 of this volume. We should recall specifically here that Article XX.g of GATT 1994 provides for a general exception allowing parties to adopt measures "relating to the conservation of exhaustible natural resources if such measures are made effective in conjunction with restrictions on domestic production or consumption". With this measure, GATT 1994 adheres to the WTO's objective of contributing to sustainable development and to the preservation of the environment. The objective is

> to ensure a balance between the right of members to take regulatory measures, including trade restrictions, to achieve legitimate policy goals (such as the conservation of natural resources) and the rights of other WTO members under basic trade rules (CITES and WTO 2015, p.8).

The incorporation of this objective in the WTO rules promotes a dialogue between the regulation of international trade and the protection of wild species. Thus, for example, the WTO and CITES have set up a system of notifications between the parties to establish reciprocal information channels regarding the measures taken by each State to meet the objective of species protection (CITES and WTO 2015, p. 5). This system promotes the objectives of both conventions, and certain reports published by WTO adjudicative bodies on dispute settlements make specific reference to CITES (Oesh 2008; WTO 1998 par. 25, 75 and 132).

So there is a link between the objectives and means of the CITES Convention and the objectives of the WTO, but at this stage this link concerns only the trading of endangered species of wild fauna and flora and their biological protection. Because of their nature, these are products that enter international trade subject to specific conditions (as do other products such as works of art).

However, what happens when the legal introduction of a product (in this case, a protected species) may pose a danger to public health? For years, the scientific community has warned of the link between public health and large-scale international trade in wildlife species (Bell et al. 2004). It is not just illegal trading that is the problem: in the area of legal trade, it is extremely difficult to carry out personalized controls of the species introduced into a country (Kolbi 2020). Therefore, control over the species imported (protected or not) focuses on the objective of preventing contagion of zoonosis.

In this case, the WTO rules applicable are Art. XX(b) GATT 1994 (on measures "necessary to protect human, animal or plant life or health") and the Agreement on the Application of Sanitary and Phytosanitary Measures (SPS).[3] Here, the dialogue between CITES and the WTO is different. In this context CITES is no longer applicable because it does not cover public health issues. However, in order to preserve public health, the WTO does incorporate rules and references to international provisions other than CITES that safeguard this common health interest and are applicable to all commerce subject to the multilateral trading system. The WTO links public health to commercial activity, not to the protection of the goods being traded or their impact in the new environment.

Therefore, the action taken must have two aims in mind: to protect the species and to protect public health, considering the two as different legal assets, each one with its own applicable rules. Both have an effect on trade, but the intended objectives are different and so are the legal bases. If we focus on the protection of public health, the legal asset being protecting by prohibiting the importation of a species is not the species itself, but public health in general. In this case the legal asset acquires a status of exception that must be added to the conditions imposed by CITES. However, in this case the exception is given not by the characteristics of the species, but by the limits generally accepted in international trade; and the legal basis is not CITES, but the regulation of international trade that incorporates the items protected by CITES as "products".

Consequently, the starting-point of our analysis is the coexistence of two different legal assets that are both in need of protection. The factor that links the preservation of endangered species of wild fauna and flora and the preservation of public health is international trade. The upshot is that, in the same legal relationship (the international commercialization of certain species) it is necessary to apply different protective legal rules, with different objectives and of different origins. This situation is also common in international legal relations.

In short, there is an area of cooperation between the WTO and CITES regarding the protection of species and their trade. In addition to this cooperation, the incorporation of specific rules regarding sustainability in the WTO regulations contributes to establishing this protective action as a common objective. However, because of the very purpose of the international trade in species (including protected species), public health may be put at risk, since the species themselves may be carriers of pathogens. Therefore, in this case, the rules that are applicable are the rules for the protection of general interests, including public health.[4] In GATT 1994 and the SPS,

[3]Significantly, although the text of the SPS Agreement refers, for example, to measures applied to protect human or animal life or health within the territory of the Member State from risks arising from diseases carried by animals, plants or products thereof (Annex A: 1 (c) SPS), at no time does it make reference to CITES. However, the text does refer to other non-WTO international instruments, such as the Codex Alimentarius Commission, the International Office of Epizootics or the International Plant Protection Convention (Annex A.3 SPS).

[4]These protective measures are not applied in a manner that would constitute a means of arbitrary or unjustifiable discrimination between Members where the same conditions prevail or a covert restriction on international trade. *Vid.* Memorandum of the World Trade Agreement on the Application of Sanitary and Phytosanitary Measures (SPS Agreement).

the WTO has specific rules for this purpose. In this way, the two objectives are promoted, although in the case of the protection of endangered species of wild fauna and flora, the protection stipulated by CITES must also be applied.

6.2.2 Protection of Endangered Species of Wild Fauna and Flora and Trade Within Regional Organizations. The European Union

In this section we analyse the relationship between the EU internal market, wildlife protection and CITES by focusing on the status of wild fauna and flora, their commercialization and, where appropriate, the protection they may enjoy. Finally, we assess the link between possible commercialization and the protection of the legal asset "public health".

The task is to establish whether wild fauna and flora come under the heading of tradable items. One of the fundamental freedoms in the Internal Market is the free movement of goods. Although there is no definition of the term 'goods' in the treaties, according to the case law of the EUCJ, it refers to all 'products which can be valued in money and which are capable, as such, of forming the subject of commercial transactions' (EUCJ 1968, p.428). For the purposes of the EU internal market, then, all species, even those falling under the scope and protection of CITES, can be considered as goods.

The Union Customs Code (EU Regulation 2013) also confirms the consideration of these elements as goods (Council Regulation 1987) as does the Common Customs Tariff. The Common Customs Tariff actually establishes a nomenclature of goods, and also includes in an Annex the species that are subject to special protection under CITES.

However, neither the consideration of these elements as goods nor their incorporation in the Customs Tariff means that they are necessarily marketable. Their trade will depend on different rules, and on the conditions and limits that these rules set; the tariff only stipulates that in the event that the products on the list can be legally imported into the EU, the tariff indicated is applicable. Trade will depend, among other things, on the limits imposed.

Wildlife species are also regulated at EU level by CITES, because the EU signed up to the Convention on 8 July 2015 as an international organization. This was made possible by two steps: the introduction of an amendment of the Convention in November 2013 that allows regional economic integration organizations to become members;[5] and the Council Decision concerning the EU's accession to CITES (Council Decision 2015). Nonetheless, the implementation of CITES regulations

[5]Prior to this amendment, only States could become parties to the Convention.

and standards in the EU dates back to 1984 (EU Regulation 1982).[6] Currently the basic regulation is the *Council Regulation (EC) No 338/97 on the protection of species of wild fauna and flora by regulating trade therein,* complemented by other regulations that develop particular aspects (Council Regulation 1997; Commission Regulation 2006).

Therefore, in the interplay between trade and biological protection the applicable European regulation is the CITES treaty. In adopting its regulations, the EU has used the proviso in Article XIV.1.a of the Convention, which allows the Parties to adopt measures that are more restrictive. This option was recognized by the EUCJ in its ruling of 23 October 2001 (EUCJ 2001, points 31 and 35). Furthermore, the Regulation itself allows Member States to adopt even more stringent measures than those that appear in its text. While setting out the EU conditions for the trading of protected species of natural origin, the Regulation is again limited to setting conditions for the trading of these species, their import and export and their transport and accommodation, and on defining the derogations.

At EU level, the trading of specially protected wild species considered as commodities involves two different stages: trading in relation to third countries (import to/export from the EU), and the movement of species within the EU market, which is governed by the principle of the free movement of goods. In the first area, import to/export from the EU, the Union Customs Code lays down (EU Regulation 2013, art. 1.1) that in these activities measures will be taken to track goods (EU Regulation 2013, Article 46) either in the form of random checks or more individualized controls. The purpose of these measures is to assess the risks arising from import-export, including the assessment and prevention of situations that could endanger public health (EU Regulation 2013, art. 5.7).

Consequently, under EU rules (European Commission. Health and consumer protection), the import of wildlife species from third countries is subject to various checks: checks on derogations, checks on compliance with guarantees, and checks on risk. The derogation checks, contained essentially in the CITES treaty and the EU Regulation implementing it, ensure that certain wildlife species whose commercialization is prohibited cannot be traded. The checks on compliance with guarantees, also provided for in the CITES treaty and in the EU Regulation implementing it, ensure that the trading of wildlife species whose commercialization is not prohibited complies with certain restrictions. Finally, risk controls are no longer regulated, but the legal basis for their application is found in the Union Customs Code.

Intra-Community movement is subject to the criteria of Regulation 338/97 itself, to the limitations implemented by the Member States in the part of the Regulation that allows this possibility, but also to the provisions of the Treaty of the Functioning of the EU (TFEU) on the free movement of goods (art. 28 and following). This guarantee of free movement within the EU has some limits, namely, the exceptions stipulated in the Article 36 TFEU. As limits to the free movement of goods in the

[6]This was incorporation by reference whereby the substantive regulation is replaced by a global reference to the regulation contained in another international convention (Andrés Sáenz de Santa María 1985, p. 38).

internal market these include the protection of health and life of humans, animals or plants. Nonetheless, the final paragraph of this article states "Such prohibitions or restrictions shall not, however, constitute a means of arbitrary discrimination or a disguised restriction on trade between Member States" (EUCJ 1976, points 15 to 17; EUCJ 2001; EUCJ 2015a, point 33; EUCJ 2019, point 29). As the EUCJ has pointed out with regard to these limitations, "a measure having equivalent effect to a quantitative restriction on imports can be justified, for example, on grounds of the protection of the health and life of humans, within the meaning of that article, only if that measure is appropriate for securing the achievement of the objective pursued and does not go beyond what is necessary in order to attain it" (EUCJ 2019, point 29). These are not, therefore, unconditional restrictions, but they must make the principle of the free movement of goods compatible with the principle of the protection sought. In the case of protected wildlife species, the balance in the internal market must ensure all three: the protection of the free movement of goods, the protection of public health, and the protection of the species.

At the legal level, then, a situation similar to the one identified in the WTO exists in the EU, albeit with the application of different rules. First, wildlife species are defined as goods for the purpose of trade. Second, there are different legal assets to protect: endangered species and public health. When these legal assets are linked by trade, the rules that apply are different. However, at EU level these rules are those of the EU: the CITES Regulation has been adapted from a EU Regulation, and the rules on public health protection are also those of the EU. Furthermore, in the case of the EU, the interpretation of the EUCJ is of particular importance.

6.3 Public Health and the International Trade in Endangered Species

In this section we focus on the relationship between public health and the international trade in protected species. The scientific community has established two types of relationship in this regard: a positive one, in which trade contributes to the development of new medicines and to scientific progress in disease control; and a negative one, in which this same trade may be the focus for the introduction of pathogens and diseases in the environments in which these species are introduced (Keesing et al. 2010; Langlois et al. 2012).

Below, we examine the legal aspects of the possible harm to public health in the country of introduction of the importation of foreign species (Scanlon 2020a), given that wildlife species can transmit diseases that affect human or animal health (Kolbi 2020; Karesh 2009). Two key concepts must be identified: the concept of public health, and the consideration of protected species as tradeable items.

6.3.1 The Concept and the Scope of Public Health

There is no single, common concept of public health at the global level. However, several definitions have been widely accepted.

Following Kemper (2015), the first one was proposed by Charles-Edward Amory Wilson (1920). For him,

> Public health is the science and art of preventing disease, prolonging life and promoting physical health and efficiency through organised community efforts for the sanitation of the environment, the control of community infections, the education of the individual in principles of personal hygiene, the organization of medical and nursing service for the early diagnosis and preventive treatment of disease, and the development of social machinery which will ensure to every individual in the community a standard of living adequate for the maintenance of health (Kemper 2015).

The second definition is due to Donald Acheson, who characterized public health in 1988 as

> the science and art of preventing disease, prolonging life and promoting, protecting and improving health through the organized efforts of society (WHO 1998, p. 3).

Two other more recent concepts, which perhaps are more in line with the current situation, also require consideration: global health and One World One Health. According to the WHO, the term "global health" refers "to the transnational impacts of globalization upon health determinants and health problems which are the beyond the control of individual nations" (Smith 2006). For its part, the concept "One World One Health" is the result of the joint efforts of four international institutions, the Food and Agriculture Organization (FAO), the World Organization for Animal Health (OIE), the World Health Organization (WHO), United Nations Children's Fund (UNICEF), supported by the World Bank and the UN System Influenza Coordinator. The objective of this strategy is to design actions to address the global consequences of emerging infectious diseases, in part produced by the globalization of the wildlife trade (FAO et al. 2008). This strategy is described in several documents, and in the field that concerns us (specifically, the relationship between wildlife, species protection, species commercialization and public health) (One Health Commission 2004), the call made to civil society and scientific communities in point 7 of the Manhattan Principles on One World One Health is especially relevant:[7] to

> reduce the demand for and better regulate the international live wildlife and bush meat trade not only to protect wildlife populations but to lessen the risks of disease movement, cross-species transmission, and the development of novel pathogen-host relationships. The costs

[7] The Manhattan Principles were drawn up at a meeting of "Health experts from around the world for a symposium focused on the current and potential movements of diseases among human, domestic animal, and wildlife populations organized by the Wildlife Conservation Society and hosted by The Rockefeller University. Using case studies on Ebola, Avian Influenza, and Chronic Wasting Disease as examples, the assembled expert panelists delineated priorities for an international, interdisciplinary approach for combating threats to the health of life on Earth".

of this worldwide trade in terms of impacts on public health, agriculture and conservation are enormous, and the global community must address this trade as the real threat it is to global socioeconomic security.

All these concepts share a common goal: the prevention and control of the spread of diseases in order to safeguard people, animals, and species in general. It is essential to identify the elements that contribute to the spread of disease (for example, wildlife species) which require particular attention from national and international institutions in order to limit or prevent the damage they may cause. In this context, the concept of EcoHealth may be also relevant.

6.3.2 The Relationship Between Public Health and International Trade

The relationship between public health and international trade is a link that affects any item placed on the market. Following the serious problems of zoonosis in recent years, international trade in wild species and its impact on public health has received particular attention. In general, for the protection of public health, both the WTO and the EU incorporate rules which aim to make international trade compatible with the protection of higher interests.

The approach to public health from an international trade perspective can be analysed from several different angles. On the one hand, the WTO can consider public health sector as a trade or service element in itself, given its role as a creator of intellectual property, as a provider of essential medicines and public care (Labonte and Sanger 2006, p. 739), and as a supplier of cross-border health services (Labonte and Sanger 2006, p. 740). On the other hand, the reference to public health in the WTO is also considered as a higher interest in need of protection, and is designed to limit the international commercialization of products and thus preserve the public health of the importing State. Unlike the previous approach, the aim is not to promote international trade in products and services (i.e., of goods linked to public health), but to limit it, applying restrictions on any product or service that may endanger the public health of the receiving country. In exercising this power, States must avoid creating "unnecessary obstacles to international trade", and are obliged to implement "any alternative measure that is less trade-restrictive" (WTO 2002, p. 32).

In the context of the WTO, the most important provisions are Article XX.b of GATT 1994 and two agreements: the Agreement on Technical Barriers to Trade (TBT Agreement) (WTO 2002, p. 32) and the SPS Agreement. These provisions allow Member States to take measures to protect public health provided that these measures are not arbitrary or involve unjustifiable discrimination with the object or effect of hindering international trade.

In the context of the WTO, therefore, measures relating to public health include action aimed exclusively at protecting the population (human, animal or plant) from external diseases. These generalist and protectionist measures apply to any product that may harm, actually or potentially, a State's public health. They therefore include species from the wild or species with special protection.

The EU is the other international organization that should be considered in this chapter in relation to international trade. Here, public health also plays an important role in the limits on commercialization. In the TFEU, and in the area of the internal market, the consideration of public health is an exception that makes it possible to limit the freedoms that are the basis of this market. The imposition of this limitation is in the context of the free movement of goods (TFEU, art. 36), the free movement of persons (EC Directive 2004), the right to freedom of establishment (TFEU, art. 52) and the freedom to provide services (TFEU, art. 62). However, none of these provisions of the original law contain any definition of the concept of public health. The closest definition is to be found in the secondary legislation, namely Article 29 of Directive 2004/38, which governs the right of residence of nationals of EU Member States and offers a definition by reference. According to this Article 29: 1.

> The only diseases justifying measures restricting freedom of movement shall be the diseases with epidemic potential as defined by the relevant instruments of the World Health Organization and other infectious diseases or contagious parasitic diseases if they are the subject of protection provisions applying to nationals of the host Member State (EC Directive 2004).

However, this lack of definition does not imply, as it does in the case of the WTO, that there are no criteria for its application. With the aim of linking freedom of trade and the protection of general interests, several criteria may be applied. As the EUCJ points out, health is an essential objective of the TFEU, and "human health and life rank foremost among the assets and interests protected by the Treaty" (EUCJ 2017, point 42). It is undoubtedly within this priority that the TFEU and the EUCJ justify their acceptance of derogations from the fundamental freedoms of the internal market on imperative grounds of the general interest (EUCJ 2015b, point 58), such as the protection of public health (EUCJ 2018, point 82).

As for the content of the derogation, the content of the concept of public health is not unique; Member States must define it, and can decide the level of protection they wish to ensure and the way to achieve this goal (EUCJ 2017, point 45). As the Court stresses, "health and the life of humans rank foremost among the property or interests protected by Article 36 and it is for the Member States, within the limits imposed by the Treaty, to decide what degree of protection they intend to assure and in particular how strict the checks to be carried out are to be," (EUCJ 1989, point 17; EUCJ 2003, point 103). However, this freedom is not absolute; as the EUCJ has stated, in this protection States must respect the limits imposed by EU law itself (EUCJ 1991a, point 26 and point 27; EUCJ 1992, point 11; EUCJ 1991b, point 53).

All this has several consequences. Firstly, the content and level of protection is not unique, as it may differ from one State to another (EUCJ 2016, point 39; EUCJ 2015c, point 118 and important case law cited). Secondly, despite these differences, States must respect the principle of proportionality in adopting limits arising from issues of public health (EUCJ 2015b, point 59). Thirdly, States may under no circumstances adopt measures that go beyond what is necessary to achieve the protection sought (EUCJ 2019, point 29; (EUCJ 2015a, point 33; EUCJ 1976, points

15–17). They may limit the exercise of the freedoms of the internal market in their regulations only if "such national legislation is appropriate for ensuring attainment of the objective relied upon only if it genuinely reflects a concern to attain that objective in a consistent and systematic manner" (EUCJ 2015b, point 65; EUCJ 2014, point 39; EUCJ 2013, point 47; EUCJ 2010, point 94). Finally, derogations on public health grounds will be compatible with the EU freedoms

> only to the extent to which they are necessary for the effective protection of the health and life of humans. National rules or practices do not qualify for a derogation under Article 36 if the health and life of humans can be protected as effectively using measures which are less restrictive of intra Community trade (EUCJ 1991a, point 27; EUCJ 2003, 1994, point 17; EUCJ 1989).

6.4 CITES: An Instrument for the Protection of Public Health?

As noted above, CITES is a convention that aims to protect endangered species. An analysis of its contents suggests that the protection of public health in a country which imports species from other countries does not fall under its competence (Scanlon 2020a): in fact, these issues are regulated either by the national laws on plant protection and public health protection,[8] or by rules of international origin. Both these legal sources, national and international, coexist in the same legal order. Therefore, it is worth asking whether CITES is the right setting for a consideration of these issues. Scanlon (Scanlon 2020b) favours their inclusion in the Convention, arguing that one of the options to consider would be

> to amend CITES to include public health criteria for listing species, possibly via a new Appendix, create a new Committee on Public Health, and oblige Management Authorities to take account of public health issues before issuing any certificate or permit (Scanlon 2020b).

However, I believe that certain reflections are in order. Firstly, it would indeed be appropriate to consider the link between public health and the trading of species in the Convention's text. However, the question arises as to what scope this consideration should have, especially whether it should entail a regulation of this link or whether a mere reference to internal or international texts on international trade would be sufficient.

The text of CITES does not incorporate phytosanitary or zoonotic issues, which are the ones most closely linked to public health. Furthermore, CITES does not regulate either the conditions or the effects of trade involving the import or export of species. So it must be determined whether the fact that the mobility of protected species has consequences for public health justifies its regulation in the context of

[8] In Spain, Law 33/2011, of 4 October, General Law on Public Health, Official State Gazette (B.O.E.) number 240, of 5 May 2011.

biological protection, or whether, on the contrary, it should be treated as a trading activity like any other.

Deciding on this point requires a contextualization of each area and each boundary in the appropriate environment. CITES has a specific purpose, but the relationship with public health is circumstantial; the protection of public health is not its object, even though, obviously, this does not mean that CITES is alien to such protection.

Moreover, the texts regulating international trade do not aim to protect species, but to establish rules for the international commerce of goods, including wildlife.[9] These rules set limits on the trading of goods when they aim to protect certain higher and collective interests, including public health. For this reason, because the connection of the protection of public health is to trade and not to biological protection, the appropriate setting for its regulation is that of the texts regulating international trade.

The development of this exception (the exception of commercialization because of public health) in other texts may not only be inappropriate but may lead to duplication, contradictions and unnecessary regulation. It is true that the solution of this problem entails the application of different regulations, of a different origin and with different objectives, to the same legal relationship, each. However, applied together they do provide the intended protection and each one limits its scope to the object for which it has been designed. In fact, this is the standard solution in the international trade of any product or merchandise.

This solution, that from my point of view is compatible with the One Health approach, maintains on the one hand the principle of unity between the conditions for international trade and exceptions to this trade when they may affect public health, and on the other hand the correlation between the trading and the protection of endangered species. It is simply a matter of applying the various appropriate rules when they need to be applied together, with the possibility of cross-referencing the texts where necessary.

Consequently, to the question posed in the title of this section, the answer should be that CITES is not (and perhaps should not be) an instrument for the protection of public health. It is an instrument for the protection of species, and exceptions to trade on phytosanitary or zoonotic grounds should be found in the trade regulations (and their references to other international rules, such as those on animal health and zoonosis of the International Office of Epizootics)—not in regulations for the protection of species.

Perhaps one option for establishing the links between these areas could be to organize a codification of the regulations existing at international level that are applicable for the trading of protected species. This would provide an overview that could link together aspects which in practice are already in place, but which from the regulatory point of view are dispersed: international trade, biological

[9]On this subject *vid.* especially Chap. 5: Conservation and sustainable use of biodiversity in the international regulation of trade in goods.

protection, limits on trading to ensure this protection, and general limits on trading to safeguard public health.

6.5 Conclusion

Globalization has affected all areas of trade between States, and species from the wild are no strangers to this trade. Regulations exist governing the protection of wildlife species and limiting their commercialization, but the legal movement of species is part of international trade. This movement involves elements that deserve special attention: biological protection, which entails that the trading of certain species must be limited in order to guarantee their survival; and the possible impact on public health of the importation of these species, given that they may carry pathogens and introduce them into their new environment.

The international regulations and conventions governing international trade should pay particular attention to the trade in these products. Even if the goods in question are traded, they must be considered particularly sensitive not just because of their own nature but because of the consequences that their incorporation into a given environment may entail. The vulnerability of public health and its effect on the international wildlife trade has already been demonstrated by the experience of pandemics and is well known in the scientific world. It is necessary to make appropriate connections between the various legal instruments applicable to the trade in these species in order to heighten the effectiveness of the protection that already exists at a regulatory level.

I am not convinced of the need for further legislation. With better application and interpretation and, where appropriate, codification, it might be possible to access all the relevant regulations, compare them and, if necessary, adapt them so that they do indeed provide the protection intended.

References

Andrés Sáenz de Santa María, M.ªP. (1985). La incorporación por referencia en el Derecho de los tratados. *Revista Española de Derecho Internacional, 37*(1), 7-40.

Bell, D., et al. (2004, June 3). Animal origins of SARS (Coronavirus): Possible links with the international trade in small carnivors. *The Royal Society*. Retrieved September 25, 2020, from https://royalsocietypublishing.org/doi/10.1098/rstb.2004.1492

CITES and WTO. (2015). *Enhancing cooperation for sustainable development*, Switzerland. Retrieved September 25, 2020, from https://www.wto.org/english/tratop_e/devel_e/citesprogr15_e.htm

Commission Regulation. (2006). *(EC) No 865/2006 laying down detailed rules concerning the implementation of Council Regulation (EC) No 338/97 on the protection of species of wild fauna and flora by regulating trade therein*. Retrieved September 25, 2020, from https://eur-lex.europa.eu/legal-content/EN/ALL/?uri=CELEX%3A32006R0865

Convention on International Trade in Endangered Species of Wild Fauna and Flora (CITES) (1973). Retrieved September 25, 2020, from https://www.cites.org/eng/disc/text.php

Council Decision. (2015). *(EU) 2015/451, concerning the accession of the European Union to the Convention on international trade in endangered species of wild fauna and flora (CITES) adopted on 6 March 2015.* Official Journal L75 of 19 March 2015. Retrieved September 25, 2020, from https://eur-lex.europa.eu/legal-content/EN/TXT/?uri=CELEX:32015D0451

Council Regulation. (1982). (EEC) No 3626/82 of 3 December 1982 on the implementation in the Community of the Convention on international trade in endangered species of wild fauna and flora. *Official Journal L 384, 31/12/1982 P. 0001 – 0061.* Retrieved September 25, 2020, from https://eur-lex.europa.eu/legal-content/EN/TXT/?uri=CELEX%3A31982R3626

Council Regulation. (1987). *(EEC) No 2658/87 of 23 July 1987 on the tariff and statistical nomenclature and on the common customs tariff. Official journal L256 of 7 September 1987.* For the consolidate version vid. Retrieved September 24, 2020, from https://eur-lex.europa.eu/legal-content/EN/TXT/?uri=CELEX:01987R2658-20200505

Council Regulation. (1997). (EC) No 338/97 on the protection of species of wild fauna and flora by regulating trade therein. *Official Journal L61 of 3 March 1997. This* Regulation has been changed several times. Consolidate version is available in Retrieved September 24, 2020, from https://eur-lex.europa.eu/legal-content/ES/TXT/PDF/?uri=CELEX:01997R0338-20200101&from=EN

European Community (EC) Directive. (2004). *Directive 2004/38 of 29 April 2004 on the right of citizens of the Union and their family members to move and reside freely within the territory of the Member States amending Regulation (EEC) No 1612/68 and repealing Directives 64/221/ EEC, 68/360/EEC, 72/194/EEC, 73/148/EEC, 75/34/EEC, 75/35/EEC, 90/364/EEC, 90/365/ EEC and 93/96/EEC.* Retrieved September 25, 2020, from https://eur-lex.europa.eu/legal-content/EN/TXT/?uri=CELEX:32004L0038

FAO et al. (2008). *Contributing to One World, One Health. A Strategic Framework for Reducing Risks of Infectious Diseases at the Animal-Human-Ecosystems Interface.* Retrieved September 25, 2020, from https://www.oie.int/doc/ged/D5720.PDF

Karesh, W. B. (2009). One World- One Health. *Clinical Medicine*, London, July. Retrieved August 23, 2020, from https://www.researchgate.net/publication/26697450_One_world-one health

Keesing, F., et al. (2010). Impacts of biodiversity on the emergence and transmission of infectious diseases. *Nature, 468*, 647–648. Retrieved August 23, 2020, from https://www.researchgate.net/publication/49651538_Impacts_of_biodiversity_on_the_emergence_and_transmission_of_infectious_diseases

Kemper, S. (2015, June 2). C-E.A. Winslow, who launched public health at Yale a Century ago, still influential today. *Yale News*. Retrieved August 23, 2020, from https://news.yale.edu/2015/06/02/public-health-giant-c-ea-winslow-who-launched-public-health-yale-century-ago-still-influe

Kolbi, J. (2020, May 7). To prevent the next pandemic, it's the legal wildlife trade we should worry about? *Animals Wildlife Watch*. Retrieved August 22, 2020, from https://www.nationalgeographic.com/animals/2020/05/to-prevent-next-pandemic-focus-on-legal-wildlife-trade/

Labonte, R., & Sanger, M. (2006, September). Glossary on the World Trade Organization and public health (part 2). *Journal of Epidemiology and Community Health*. Retrieved August 26, 2020, from https://www.ncbi.nlm.nih.gov/pmc/articles/PMC2566018/

Langlois, E., et al. (2012). Towards a better integration of global health and biodiversity in the new sustainable development goals beyond Rio+20. *Ecohealth, 12*, 381–385. Retrieved August 19, 2020, from https://pubmed.ncbi.nlm.nih.gov/23001330/

Law (Spain) 33/2011, of 4 October, General Law on Public Health, Official State Gazette (B.O.E.) number 240, of 5 May 2011. Retrieved August 20, 2020, from https://www.boe.es/buscar/act.php?id=BOE-A-2011-15623#:~:text=La%20ley%20reconoce%20el%20derecho,previstas%20en%20la%20normativa%20vigente

Oesh, M. (2008). *US-Shrimp Case, Max Planck Encyclopedia of Public International Law*. Oxford: Oxford University Press. Retrieved August 17, 2020, from https://opil.ouplaw.com/view/10.1093/law:epil/9780199231690/law-9780199231690-e1699

One Health Commission. (2004). *The Manhattan Principles on one World one Health*. Retrieved September 19, 2019, from https://www.onehealthcommission.org/index.cfm/37526/75802/twelve_manhattan_principles_formulated. For the update version vid. "The 2019 Berlin Principles on One Health" https://oneworldonehealth.wcs.org/About-Us/Mission/The-2019-Berlin-Principles-on-One-Health.aspx

One World, One Health. Retrieved August 25, 2020, from http://www.oneworldonehealth.org/

Regulation. (1982). (EEC) No 3626/82 of 3 December 1982 on the implementation in the Community of the Convention on international trade in endangered species of wild fauna and flora, and amendments. *Consolidated version*. Retrieved September 25, 2020, from https://ec.europa.eu/inea/sites/inea/files/download/legal_framework/delegated_regulation12682012.pdf

Regulation. (2013). (EU) No 952/2013 of the European Parliament and of the Council of 9 October 2013 laying down the Union Customs Code. *Official Journal L260 of 10 October 2013*. For the consolidate version *vid*. Retrieved September 9, 2020, and September 24, 2020, from https://eur-lex.europa.eu/legal-content/EN/TXT/?uri=CELEX:02013R0952-20200101

Scanlon, J. E. (2020a). Wildlife Trade, Origins of COVID-19, and Preventing Future Pandemics. U.S. Congressional International Conservation Caucus Virtual Caucus. *Hearing "Wildlife Trade, Origins of COVID-19, and Preventing Future Pandemics"*, 15 April. Retrieved August 25, 2020, from https://www.linkedin.com/pulse/wildlife-trade-origins-covid-19-preventing-future-uk-scanlon-ao/

Scanlon, J. E. (2020b). *A fresh look at Global wildlife trade law: Can CITES help prevent pandemics?* IUCN Commission on Environmental Law; Webinar Series - 18 May. Retrieved August 2, 2020, from https://www.linkedin.com/pulse/fresh-look-global-wildlife-trade-law-can-cites-help-scanlon-ao/

Smith, B. J. (2006). WHO Health Promotion Glossary: New Terms. *Health Promotion International Advance Access*, September 7, Oxford University Press. Retrieved September 10, 2020, from https://www.who.int/healthpromotion/about/HP%20Glossay%20in%20HPI.pdf?ua=1

Treaty on the Functioning of the European Union (TFEU). *Consolidated version*. Retrieved September 24, 2020, from https://eur-lex.europa.eu/legal-content/EN/TXT/?uri=CELEX%3A12012E%2FTXT

WHO. (1998) *Health Promotional Glossary, Geneva*. Retrieved August 24, 2020, from https://www.who.int/healthpromotion/about/HP%20Glossay%20in%20HPI.pdf?ua=1

WTO. (1998). Appellate Body report adopted on 6 November 1998 in WTO, *US - Shrimp*, Report of the Appellate Body, WT/DS/AB/R, 12 October. Retrieved August 26, 2020, from https://www.wto.org/english/tratop_e/dispu_e/cases_e/ds58_e.htm

WTO. (2002). *Agreements and public health: a joint study by the WHO and WTO secretariat*. Retrieved September 20, 2020, from https://apps.who.int/iris/handle/10665/42535

Case Law

EU Court of Justice. (EUCJ). (1968). Judgment of 10 December 1968, *Commission v. Italy*, (7/68, EU:C:1968:51)

EU Court of Justice. (EUCJ). (1976). Judgement of 20 May 1976, *Adriaan de Peijper, Managing Director of Centrafarm BV.*, (104/75, EU:C:1976:67)

EU Court of Justice. (EUCJ). (1989). Judgement of 7 March 1989, *Heinz Schumacher, v. Frankfurt am Main 1 and Hauptzollamt (Principal Customs Office) Frankfurt am Main-Ost*, (215/87, ECLI:EU:C:1989:111)

EU Court of Justice. (EUCJ). (1991a). Judgement of 16 April 1991, *Freistaat Bayern v Eurim-Pharm GmbH,* (C-347/89, ECLI:EU:C:1991:148)
EU Court of Justice. (EUCJ). (1991b). Judgement of 21 March 1991 *Criminal proceedings against Jean-Marie Delattre*, (C-369/88, ECLI:EU:C:1991:137)
EU Court of Justice. (EUCJ). (1992). Judgement *of 8 April 1992, Commission of the European Communities v Federal Republic of Germany,* (C-62/90, ECLI:EU:C:1992:169)
EU Court of Justice. (EUCJ). (1994). Judgement of 10 November 1994, *Lucien Ortscheit GmbH v Eurim-Pharm Arzneimittel GmbH,* (C-320/93, ECLI:EU:C:1994:379)
EU Court of Justice. (EUCJ). (2001). Judgement of 23 October 2001, Xavier Tridon, and Fédération départementale des chasseurs de l'Isère, v. Fédération Rhône-Alpes de protection de la nature (France), Isère section, (C-510/99, ECLI:EU:C:2001:559)
EU Court of Justice. (EUCJ). (2003). Judgement of 11 December 2003, *Deutscher Apothekerverband eV and 0800 DocMorris NV, Jacques Waterval,* (C-322/01, ECLI:EU:C:2003:664)
EU Court of Justice. (EUCJ). (2010). Judgement of 1 June 2010, *Blanco Pérez and Chao Gómez v Consejería de Salud y Servicios Sanitarios*, (C-570/07 and C-571/07, EU:C:2010:300)
EU Court of Justice. (EUCJ). (2013). Judgement of *26 September 2013, Ottica New Line di Accardi Vincenzo v Comune di Campobello di Mazara, (*C-539/11, EU:C:2013:591)
EU Court of Justice. (EUCJ). (2014). Judgement of 13 February 2014, *Susanne Sokoll-Seebacher,* (C-367/12, EU:C:2014:68)
EU Court of Justice. (EUCJ). (2015a). Judgement of 23 December 2015, Scotch Whisky Association and Others v The Lord Advocate and The Advocate General for Scotland (C-333/14, EU:C:2015:845)
EU Court of Justice. (EUCJ). (2015b). Judgement of 23 December 2015, Gebhart Hiebler v. Walter Schlagbauer, (C-293/14 ECLI:EU:C:2015:843)
EU Court of Justice. (EUCJ). (2015c). Judgement of 12 November 2015, Valev Visnapuu v. Kihlakunnansyyttäjä y Suomen valtio – Tullihallitus, (C-198/14, EU:C:2015:751)
EU Court of Justice. (EUCJ). (2016). Judgement of 19 October 2016 Deutsche Parkinson Vereinigung eV v Zentrale zur Bekämpfung unlauteren Wettbewerbs eV, (C-148/15 ECLI:EU:C:2016:776)
EU Court of Justice. (EUCJ). (2017). Judgement of 20 December 2017, Sabine Simma Federspiel v Autonomous Province of Bolzano, and Equitalia Nord SpA, (C-419/16, ECLI:EU:C:2017:997)
EU Court of Justice. (EUCJ). (2018). Judgement of 26 September 2018, Criminal proceedings Van Gennip BVBA, Antonius Johannes Maria ten Velde, Original BVBA, Antonius Cornelius Ignatius Maria van der Schoot (C-137/17, ECLI:EU:C:2018:771)
EU Court of Justice. (EUCJ). (2019). Judgement of 3 July 2019, Delfarma sp. z o.o. v Prezes Urzędu Rejestracji Produktów Leczniczych, Wyrobów Medycznych i Produktów Biobójczych. (C-387/18, ECLI:EU:C:2019:556)

Chapter 7
Access to and Benefit-Sharing of Genetic Resources: Rationale, Review and Prospects

Xavier Pons Rafols

Contents

7.1 Introduction 120
7.2 Access to Natural Resources and the Fair and Equitable Sharing of the Benefits Arising Out of Their Use in International Law 122
7.3 Access to Genetic Resources and the Sharing of Benefits Arising from Their Use in the International Biodiversity Regime 125
7.4 The Specific Multilateral System of Access and Benefit-Sharing of Genetic Resources for Food and Agriculture 127
7.5 The Impact on the ABS System of Digital Information on Genetic Sequences 129
7.6 Access and Benefit-Sharing in Relation to Marine Genetic Resources in Areas Beyond the Jurisdiction of States 131
7.7 Conclusion 134
References 135

Abstract This chapter examines the rationale for the international regime governing access and benefit-sharing (ABS) in relation to genetic resources, offering an overview of the regime and looking at future prospects. After emphasizing the value of genetic resources for biological diversity, the chapter turns to the rationale for regulating access to natural resources and the fair and equitable sharing of benefits arising from their use, from the perspective of international law. This approach leads on to an analysis of the scope and limits of the ABS system in the international biodiversity regime and the specific multilateral ABS regime on plant genetic resources for food and agriculture. In a prospective review, the analysis looks at the digitization of genetic sequence information, which is currently one of the main challenges facing the internationally established ABS mechanism, and then explores ongoing work on an ABS mechanisms for a future agreement on marine biodiversity in areas beyond national jurisdiction.

X. Pons Rafols (✉)
Universitat de Barcelona, Faculty of Law, Barcelona, Spain
e-mail: xpons@ub.edu

M. Campins Eritja, T. Fajardo del Castillo (eds.), *Biological Diversity and International Law*, https://doi.org/10.1007/978-3-030-72961-5_7

7.1 Introduction

In recent decades, the conservation of biological diversity as an expression of the variety of life forms—plants, animals and microorganisms—and of the ecosystems and ecological processes that exist on our planet has become as one of the foremost concerns of the international community. This development reflects a collective worry over the sharp reduction in biodiversity and it recognizes both the vital importance and intrinsic value of such diversity. It also gave rise to the Convention on Biological Diversity (CBD) of 5 June 1992, which entered into force on 29 December 1993, and the international legal regime established around the CBD.

The foundation of biodiversity lies in genetic material, i.e. any material of plant, animal, microbial or other origin containing functional units of heredity. Where genetic material has actual or potential value, i.e. it can be used for economic or commercial purposes, it amounts to genetic resources that generally fall under the category of biological resources (Article 2 CBD). In addition to any direct benefits from the use and consumption of genetic resources, the economic benefits that can be gained from these resources have only grown, so that their contributions to agricultural and agri-food innovations, pharmaceutical and cosmetic products, and numerous science- or biotechnology-based industrial activities are now enormous (Kate and Laid 2000). This has led the Food and Agriculture Organization (FAO) of the United Nations to promote the International Treaty on Plant Genetic Resources for Food and Agriculture (ITPGRFA), which was adopted by the FAO Conference on 3 November 2001 and has been in force since 29 June 2004.

Advances in life sciences and biotechnology have further increased the interest in and value of the components of biodiversity and genetic resources. Consequently, greater international attention is being paid to the fair and equitable sharing of benefits that arise from bioprospecting and scientific research and development in a field that has a transnational component with implications for international law. In this regard, scientific progress on the molecular structure of cells and continuing developments in genetic engineering have multiplied exponentially since the description of deoxyribonucleic acid (DNA) in 1953 and the development of genetically modified organisms (GMOs) and animal cloning through the nuclear transfer of adult cells in the 1980s and 1990s. The latest stages in this frontier science have witnessed the mounting speed and mastery of genome sequencing, as well as recent new gene-editing techniques such as CRISPR/Cas9, which herald the field's most powerful prospects for the future.

In any case, a double asymmetry lies at the heart of the international debate on genetic resources and it is reflected not only in the difficulties posed by the establishment of effective mechanisms for access to and use of these resources to bring about a fair and equitable sharing of the benefits, but also in the limited practical outcome, in monetary and non-monetary terms, of the mechanisms and systems that currently exist at the international level. On one hand, it is an asymmetry between States with greater biological diversity and greater genetic resources, such as megadiverse States, and States which, for geographical, historical, climatic and biological reasons, have fewer genetic resources in their territory. On the other hand, it is an asymmetry between States that have economic resources and research and technological capacity to take advantage of genetic resources, both in research and in their industrial and commercial uses, and States that lack such scientific and technological capacities. Broadly speaking, the axis that defines this double asymmetry is also the axis that exists between developed and developing States and it reflects the inequities of the contemporary world.

At the same time, the unequal distribution of genetic resources, their progressive deterioration and the resulting loss of biodiversity, as well as the need for their conservation, have generated international action to safeguard their preservation and sustainability and guarantee equitable access and distribution of the benefits derived from their use. For a long time, access to genetic resources was open and constituted an inequitable exploitation in relation to the States providing them, since it was understood that they should not be paid. In the 1980s, it began to be recognized that the States that possess many of these resources should be able to obtain a fair and equitable share of the benefits and profits derived from their use, if only to make the obligation to protect species and their ecosystems more attractive, while also keeping the door open to scientific innovation and development. The demands of developing States were given concrete expression in the CBD which, for the first time, recognizes the sovereignty of States over the genetic resources in their territory. Furthermore, the equitable sharing of benefits from the use of genetic resources has had an impact on the rights and traditional knowledge of indigenous and local communities within States.[1]

In other words, the CBD places value on access to genetic resources and on their international importance, as well as on the need for fair and equitable sharing—including at the international level—of the benefits derived from their use, which has ultimately become regulated under the international biodiversity regime. This is why one of the CBD's three objectives, in addition to the conservation of biological diversity and the sustainable use of its components, is the fair and equitable sharing of the benefits arising out of the utilization of genetic resources, including by appropriate access to genetic resources and by appropriate transfer of relevant technologies, taking into account all rights over those resources and to technologies, and by appropriate funding (Article 1 CBD). The result was the establishment of an access and benefit-sharing (ABS) system. Specifically, the ABS dimension has been developed through the Nagoya Protocol on Access to Genetic Resources and the Fair and Equitable Sharing of Benefits Arising from their Utilization to the Convention on Biological Diversity (NP), which was adopted on 29 October 2010 and has been in force since 12 October 2014.

---

[1]Attention must also be paid to respecting and maintaining the knowledge, innovations and practices of indigenous and local communities (ILCs). In this respect, there is an important dimension of intra-state benefit-sharing that is covered by provisions of international law. However, this study deals exclusively with the dimension of inter-state benefit-sharing.

The focus of this chapter is on access and benefit-sharing. Following the introduction, the next section addresses access to natural resources and the fair and equitable sharing of benefits from their use in a more general manner. The third section then formulates a broad overview of the scope and limits of the ABS system within the international biodiversity regime, which pivots on the CBD and especially on the NP. The fourth section analyses the specific multilateral ABS regime established in ITPGRFA and the work now underway to strengthen and reform it. Subsequently, the chapter analyses the digital sequence information (DSI) of genetic resources, which is currently one of the major challenges facing the entire internationally established ABS mechanism. Finally, the chapter concludes with a review of the work underway towards a future agreement on marine biodiversity in areas beyond national jurisdiction (ABNJ) and the possibility of establishing an ABS mechanism for marine genetic resources.

7.2 Access to Natural Resources and the Fair and Equitable Sharing of the Benefits Arising Out of Their Use in International Law

Since the 1950s, the process of decolonization has raised international concerns about development. The new States that emerged out of decolonization pushed for a new international order that would take into account their interests and their position in the international system, pursuing a strategy that culminated in 1974 with the formulation of the so-called New International Economic Order (UNGA 1974a) and the adoption of the Charter of Economic Rights and Duties of States (UNGA 1974b). Both texts set out legal principles that are intended to inspire relations between developed and developing States, to the benefit of the latter. Their approaches were guided by a two-pronged approach that has, to a large extent, converged. The first prong relates to the issue of human rights, which in turn can be viewed from three perspectives: first and most broadly, the right of all human beings to share scientific progress and its benefits (UNGA 1948, Article 27.1); second, so-called third-generation rights or solidarity rights, such as, for example, the right to development, which has been formulated by the General Assembly (UNGA 1986); and third, the rights of indigenous peoples, also proclaimed by the General Assembly (UNGA 2007) and by the 1989 Indigenous and Tribal Peoples Convention of the International Labour Organization (ILO). The second prong, in as far as it is relevant here, relates to the full sovereignty of States over their resources and natural wealth.

Ultimately, the confluence of the two prongs was subsumed within a balancing of priorities between the concerns of developing States over their own development and international concerns over environmental problems, such as the depletion of the ozone layer, climate change and the conservation of biological diversity, all of which have a greater international dimension and can therefore only be addressed through international cooperation. The direction began to change in 1987 with the Report of the World Commission on Environment and Development (the Brundtland Report). This report, which marked a first step towards a more balanced integration of the environment and development, set out an initial attempt at defining the notion of sustainable development. An expression of the confluence noted above, together with an enhanced formulation of sustainable development, came with the 1992 Rio Conference on Environment and Development, whose spirit and resulting texts revolve around the notion. This is true of the three texts adopted at the conference itself—particularly the Rio Declaration on Environment and Development (UN 1992)—and the two major international conventions adopted as part of the conference process: the United Nations Framework Convention on Climate Change (UNFCCC) and the CBD.

More specifically, it is epitomized in the CBD provision that recognizes economic and social development and poverty eradication as essential priorities and makes compliance by developing States with their obligations conditional on the effective implementation by developed States of their own commitments regarding finance and technology transfer (preamble and Article 20.4 CBD). Accordingly, the notion of sustainable development has become a new paradigm that encompasses various other fundamental principles for the promotion and achievement of sustainable development, such as the principle of common but differentiated responsibilities.

In relation to the sovereignty of States over their natural resources, there are numerous international texts that reflect the notion of States' full or permanent sovereignty over their natural wealth and resources, such as Resolution 1803 (XVII) (UNGA 1962) or Article 1.2 of the 1966 International Covenant on Civil and Political Rights and the 1966 International Convention on Economic, Social and Cultural Rights. This approach was later reiterated in Article 12 of the Charter of Economic Rights and Duties of States (UNGA 1974b) and in Article 1.2 of the Declaration on the Right to Development (UNGA 1986). Although all these texts and other related ones refer to natural wealth and resources and not specifically to genetic resources, the idea of a common interest of humanity and the very concept of the Common Heritage of Mankind (CHM) are inescapably present in relation to genetic resources. In any case, both Principle 2 of the Declaration on Environment and Development and Article 3 of the CBD set out limits and conditions in relation to the use of natural resources that are related to the United Nations Charter and international law.

In a similar vein, the 1960s witnessed the emergence of the aforementioned notion of a common interest of humanity, which gave rise to the concept of CHM and which expresses the necessary solidarity among States in relation to the existing resources in certain spaces that do not fall under the sovereignty of any particular State. The concept combines the progressive humanization of international society and concerns for the development of States emerging from decolonization with the exploration and exploitation of certain areas now made possible because of scientific and technological development. The concept initially surfaced in relation to the law of the sea (UNGA 1970), and is especially reflected in Article 136 of the United Nations Convention on the Law of the Sea (UNCLOS) in relation to the area covered by the seabed and ocean floor and the subsoil thereof that lies beyond the limits of national jurisdiction. The notion of CHM is also mentioned, but not defined, in Article 11 of the Agreement Governing the Activities of States on the Moon and Other Celestial Bodies of 1979. The legal content of the notion revolves around four fundamental normative principles: the principle of non-appropriation and exclusion of sovereignty; the principle of peaceful use; the principle of freedom of access, exploration and scientific research; and the principle of rational management of resources and their equitable distribution for the benefit of all humankind, i.e. with a redistributive effect. The CHM also has an institutional dimension which has become relevant through the action, for instance, of the International Seabed Authority (ISA) which organizes, regulates and controls activities in the area on behalf of humanity as a whole. The notion, therefore, expresses the idea that the benefits derived from the exploitation of the resources and wealth of areas that are not under the jurisdiction of any State should be for the benefit of the entire international community.

Also lying behind this perspective, clearly, is the deep asymmetry between developed and developing States. However, the ambiguity of this notion in terms of international political support has led to its replacement, on one hand, by an approach to global public goods (EJIL 2012; Bouza et al. 2015) and, on the other hand, by an expansion of the core idea of a common concern of humankind and an international interest in natural resources, particularly genetic ones. In this sense, the Brundtland Report had already stressed that species and genetic variability should be understood as a common heritage, but that collective responsibility for the common heritage would not mean collective international rights to particular resources within nations (UN 1987, Chapter 6, para. 59). That is, it would not affect the concept of national sovereignty, but nor would it leave each State to rely on its own isolated efforts to protect species within its borders.

The widespread notion of ABS is therefore shaped by and closely linked to underlying trends related to solidarity and sustainable development. However, it addresses resources—in this case genetic resources—that fall under the jurisdiction of States. In this respect, the various current ABS systems fit within and are relevant to the achievement of the Sustainable Development Goals (SDGs), in particular SDG 2 (Zero Hunger), SDG 14 (Life Below Water) and SDG 15 (Life on Land). Indeed, two SDG targets specifically refer to the promotion of access to and fair and equitable sharing of benefits arising from the utilization of genetic resources and associated traditional knowledge, as internationally agreed (UNGA 2015a, Targets 2.5 and 15.6). It can therefore be said that there is a clear rationale for the right to development and sustainable development throughout the ABS regime (Broggiato et al. 2015, p. 16). Furthermore, from a human rights perspective—and from the perspective of the rights of indigenous peoples and local communities—it can also be argued that fair and equitable benefit-sharing is a general principle of international law (Morgera 2016, 2018).

7.3 Access to Genetic Resources and the Sharing of Benefits Arising from Their Use in the International Biodiversity Regime

Of the three fundamental objectives of the CBD, the third one concerns access and benefit-sharing. Just as fundamental as the other two closely related objectives, it is, as noted earlier, specifically developed in Article 1 CBD, which sets out some of the means for its achievement. These are the real cornerstone of the CBD. This also implies that there may be other means to achieve the objective and that the flow of benefits is initially conceived as going from developed States to developing States, to which the relevant technologies and appropriate funding should be transferred (Pérez Salom 2002, p. 124). The relevant operation carried out by the CBD is, on one hand, to reconcile the principle of State sovereignty over genetic resources, which is formulated for the first time (Article 3), with the common interest of humankind (Preamble); and, on the other hand, to reconcile the regulation of access to genetic resources with the necessary fair and equitable sharing of the benefits arising from their use for the States under whose jurisdiction the resources are located.

This is a conciliatory operation that also establishes the right of a State to regulate access to genetic resources under its jurisdiction and the dual obligation to facilitate access and not to impose restrictions contrary to the objectives of the CBD. Based on the recognition of the sovereign rights of States over their natural resources, the CBD clearly states that the power to regulate access to genetic resources lies with national governments and is subject to national legislation (Article 15.1 CBD). Consequently, it is established that access will be subject to the prior informed consent (PIC) of the provider country, whether this is the State of origin of the resources or another State that has acquired them in accordance with the CBD; and also that, when access is granted, it will be on mutually agreed terms (MAT), i.e. by the conclusion—for each ABS transaction—of a private law contract between individual providers and users that regulates access, the conditions thereof and the corresponding fair and equitable benefit-sharing. To complete the scheme, the CBD also provides that each Party shall take measures, on mutually agreed terms, to share fairly and equitably the results of research and the benefits arising from the commercial and other utilization of genetic resources with the Party providing such resources (Article 15.7 CBD).

Despite the provisions of the CBD, however, the need to develop more clearly the ABS system to which the Convention refers led, in 2002, to the adoption of the Bonn Guidelines on Access to Genetic Resources and Fair and Equitable Sharing of the Benefits Arising out of their Utilization at CoP 6 (CBD 2002). The new guidelines, which are aimed at users and providers of genetic resources, identify the basic elements of both PIC and MAT. They provide guidance to States on the policy, legislative and administrative measures to be adopted on ABS, as well as mechanisms for fair and equitable benefit-sharing, including an indicative list of examples of monetary and non-monetary benefits. In any case, MAT and specific conditions should continue to be determined by the concerned Parties on a case-by-case basis.

The non-binding nature of the Bonn Guidelines led almost immediately (UN 2002) to the initiation of a process towards the development of the Nagoya Protocol (NP), which was signed in 2010 and is specifically dedicated to ABS. The starting point of the NP lies in a recognition of the public's awareness that the economic value of ecosystems and biodiversity and the need for fair and equitable sharing of their economic value with the custodians of biodiversity are both key incentives for the conservation of biological diversity and the sustainable use of its components (Morgera et al. 2010, 2014). Consequently, Article 1 of the NP reproduces almost verbatim the third objective of Article 1 of the CBD. The key difference lies in its specific focus, since the CBD's third objective is considered in the NP to be the first and most important objective, and thus to contribute to the achievement of the other two objectives relating to conservation and the sustainable use of biological diversity and its components.

As previously established in the CBD, the NP reiterates that benefit-sharing shall be carried out on MAT. Similarly, access to genetic resources shall be subject to national legislation and to the PIC of the Party providing such resources, that is, the country of origin of the resources or a Party that has acquired the resources in accordance with the Convention. This is how the bilateral ABS mechanism provided for by the NP is configured. The NP also establishes an Access and Benefit-sharing Clearing (Article 14 NP), which acts as a facilitating mechanism and as a means for sharing information, for example, on the diversity of internal legislative, policy and administrative provisions implementing ABS frameworks (Sirakaya 2019). In addition, the NP has an annex that lays out a fuller, more detailed list of monetary and non-monetary benefits. As indicated earlier, the non-monetary benefits include cooperation in scientific research and development programmes, collaboration in training and capacity-building, and the transfer of knowledge and technology, all on fairer and more favourable terms, including preferential conditions, as well as admission to *ex situ* facilities and databases.

The guidance derived from the CBD and, more specifically, the NP is therefore to implement a bilateral mechanism between the Party providing genetic resources, whether it is the State of origin or a Party that has acquired rights over the resources, and the Party seeking access to and use of the resources. This is a bilateral mechanism that has an inter-state dimension, between States. As noted earlier, however, it also takes on a domestic and private dimension, since the MAT will be established directly between the users and the owners of the resources, always under the legislation and with the PIC of the State where the resources are located, while also taking into account the interests and rights of indigenous and local communities.

In any case, the first evaluation and review of the NP's effectiveness led, in 2018, to the adoption of a decision which, while welcoming progress on implementation by the Parties, also underlined the need for further work (CBD 2018a). Indeed, the decision emphasized that many Parties had neither completed the implementation of legislative, administrative and policy measures on ABS nor established the necessary institutional arrangements. Unsatisfactorily, only 71% of the Parties had adopted legislative, administrative and policy measures on ABS. As a consequence, benchmarks were adopted in order to be able to measure progress in the future.

Finally, the NP also provides for the future possibility of implementing a global multilateral benefit-sharing mechanism to address ABS in relation to genetic resources in transboundary situations or where it is not possible to grant or obtain PIC (Article 10 NP). The option of a multilateral ABS mechanism, which has not yet been activated, would allow for benefits to be shared and used as a global fund to support the conservation of biological diversity and the sustainable use of its components at the global level; however, to create such a mechanism, further information is still required (CBD 2018b).

7.4 The Specific Multilateral System of Access and Benefit-Sharing of Genetic Resources for Food and Agriculture

Before the adoption of the CBD, the FAO Conference had already adopted the International Undertaking on Plant Genetic Resources in 1983 (FAO 1983). The objective of the Undertaking was to ensure that plant genetic resources of economic and/or social interest, particularly for agriculture, would be explored, preserved, evaluated and made available for plant breeding and scientific purposes. In other words, based on the perspective of food security and a perception that plant genetic resources are a heritage of mankind and must be accessible without restrictions, the aim was to ensure the safe conservation and promote the availability and sustainable use of plant genetic resources for present and future generations, providing a flexible framework for sharing benefits. Accordingly, it was envisaged that States should allow access to and export of plant genetic resources under their control when requested for scientific research, plant breeding or the conservation of genetic resources. This led to the establishment of the Global System on Plant Genetic Resources, which was intended to be beneficial to all participants and take full account of the rights of germplasm donors, funds and technology, as well as the obligations of recipients. In 1989 and 1991, respective FAO Conferences adopted three agreed interpretations of the International Undertaking on Plant Genetic Resources, which recognized the rights of technology and germplasm donors to be compensated for their contribution, as well as farmers' rights and the Undertaking's compatibility with the 1961 International Convention for the Protection of New Varieties of Plants (UPOV Convention).

Even so, it was still necessary to advance in a more determined manner towards a legally binding text. This process, which began in 1994, culminated in 2001 with the adoption, by the FAO Conference (FAO 2001), of ITPGRFA. ITPGRFA is a treaty that aims at the conservation and sustainable use of plant genetic resources for food and agriculture and the fair and equitable sharing of the benefits arising out of their use, in harmony with the CBD, for sustainable agriculture and food security. In other words, although the treaty deals only with certain plant genetic resources for food and agriculture, its purposes are conservation, sustainable development and ABS in relation to the genetic resources in question. More specifically, Annex I of ITPGRFA lists the 64 species of food crops and forages to which the treaty applies.

Basically, ITPGRFA has two aspects that are relevant for this chapter. The first aspect relates to the difference between the ABS system provided for in ITPGRFA and the NP's ABS system. Unlike the bilateral nature of the NP's system, which is specified in each ABS transaction, ITPGRFA provides for a sophisticated and complex Multilateral System (MLS) of ABS. This MLS applies to all genetic resources set out in Annex I, provided that they are under the management and control of the Contracting Parties and in the public domain; however, in order to achieve the maximum possible coverage, all other holders of these resources are also invited to join the MLS voluntarily. Access is granted solely for the purpose of utilization and conservation to conduct research, breeding and training related to food and agriculture, provided that such purpose does not include chemical, pharmaceutical and/or other non-food/feed industrial uses.

In addition, access shall be provided pursuant to a standard material transfer agreement (SMTA), which was designed and approved at the first meeting of ITPGRFA's Governing Body. That is, although the ABS conditions are also established by contract, in this case there is a simpler, standard form that defines them in a general way and not on a case-by-case basis. In the MLS benefits will be shared through the exchange of information, access to and transfer of technology, capacity-building, and the sharing of the benefits arising from commercialization. In the latter case, a financial contribution must also be made to a third party acting in a fiduciary capacity, i.e. the Benefit-sharing Fund (BSF) established by the Governing Body. Primarily, these resources should flow directly and indirectly to farmers in all countries, especially in developing countries, and countries with economies in transition, which conserve and sustainably utilize plant genetic resources for food and agriculture.

The second aspect concerns the compatibility of ITPGRFA with the international biodiversity regime. Article 1 of ITPGRFA already refers to its necessary harmony with the CBD. More specifically, the NP, which is subsequent to ITPGRFA, leaves room in its Article 4 for the adoption of other international agreements on ABS, provided that they are favourable and not contrary to the objectives of the CBD and the NP. In other words, if a specialized international instrument on ABS that is consistent with and does not run counter to the objectives of the CBD and NP is applicable, then the NP does not apply to the Party or Parties to the specialized instrument in relation to the specific genetic resources covered by the instrument. There is therefore full compatibility between the multilateral system of ABS that applies to plant genetic resources and the ABS system provided for in the international biodiversity regime, which is subsidiary for this type of genetic resources. This view has also been reiterated by the CoP and the CoP-MoP when inviting the application of ITPGRFA and the NP in a mutually supportive manner (CBD 2016a). On the other hand, many opinions hold the view that the post-2020 global biodiversity framework should make stronger reference to compatibility and coordination between the ABS system provided for in the NP and the multilateral ABS system established in ITPGRFA (CBD 2019; CGIAR et al. 2020).

However, now more than 15 years since ITPGRFA entered into force, the ABS component of the envisaged MLS has not been successful, and the expected monetary flows into the access user system to facilitate projects for the conservation and sustainable development of plant genetic resources in provider States have not occurred. As a result, the Governing Body has approved the creation of an *Ad hoc* Open-ended Working Group to Enhance the Functioning of the Multilateral System and increase contributions to the Benefit-sharing Fund (FAO 2013). Thus, within the Working Group, there has been a recent proposal to renegotiate the SMTA used within the framework of the MLS. The proposal has been associated, on one hand, with DSI, which has become the main outstanding issue preventing consensus (Karger et al. 2019, p. 11), and, on the other hand, with the future reform of Annex I of the treaty and the expansion of the list of food crops and forages established therein.

The *Ad Hoc* Open-Ended Working Group to Enhance the Functioning of the Multilateral System made progress along these lines in its 2019 report (FAO 2019a). However, at the last meeting of the Governing Body, no consensus could be reached on measures to strengthen the functioning of the ITPGRFA multilateral system (FAO 2019b). Similarly, no progress was made on the amendment proposal submitted by Switzerland, which the Working Group had discussed and addressed with a more fully developed amendment proposal to Annex I, that it incorporated in its own report (FAO 2019a). These proposals for amendment were intended to ensure that, beyond the food crops and forages set out in Annex I—to which ITPGRFA applies—the multilateral system should cover all other plant genetic resources for food and agriculture, even if they are not in the public domain and do not fall under the administration and control of the Contracting Parties.

Finally, the FAO Commission on Genetic Resources for Food and Agriculture (CGRFA) has also undertaken to analyse possible ABS systems in relation to other sub-sectors of genetic resources, especially animal, aquatic and forest genetic resources and microorganisms and invertebrates. In addition, the development of ABS Elements has been promoted so that such Elements—which have also been recognized under the CBD (CBD 2016c)—can serve as voluntary guidelines to facilitate in the national implementation of access and benefit-sharing in the other sub-sectors (FAO 2019b).

7.5 The Impact on the ABS System of Digital Information on Genetic Sequences

The important advances noted above in the field of biology and synthetic biology have, at the same time, positioned DSI as a major opportunity for innovation and as a critical aspect of ABS. As genetic data have become more quickly and more cheaply sequenced, the biological data produced, stored and made available in public genetic data banks have multiplied enormously, both for the development of scientific research and for commercial purposes. The digital availability of genetic data enables researchers and commercial users to gain access quickly and directly without any need for direct, physical access to the genetic material or any need to sequence the material (i.e. *in silico* or *ex silico* access). However, this is happening to the detriment of an ABS regime that seeks to ensure fair and equitable sharing of the use of the data by the States where the genetic material was initially located. In other words, the extension and generalization of DSI permit the use of genetic material without the user needing to obtain PIC from the provider State or share the benefits arising from its use through MAT. In particular, this can have a serious effect on developing States.

Debate over the issue has been going on within the bodies of the CBD and the NP since 2016. However, the first area of disagreement concerns the very conceptualization of DSI, since the various decisions that have been taken on the matter always state that the name DSI is to be used provisionally only until an alternative can be agreed (CBD 2016b, 2018c). That is, terms like genetic sequence data, genetic sequence information, genetic information, dematerialized genetic resources, *in silico* utilization or resources *in silico* (CBD 2018d; FAO 2018a) or, more straightforwardly, some combination of digital sequence information and genetic sequence data (DSI/GSD) is to be used until such time as new terminology shall be agreed (FAO 2019a).

The second area of disagreement concerns whether international instruments that apply to genetic resources also cover DSI or only cover access and conservation *in situ* and *ex situ* and therefore physical biological samples. In most cases, the generation of DSI for genetic resources requires access to a genetic resource, although it may be difficult to relate the DSI in question to the genetic resource from which it was generated. One option that would place DSI on the same level as *in situ* access would be to verify the traceability of DSI, so that even if genetic resources are not physically available, it may be possible to trace their origin and thus identify the provider country. In light of the evolution of science and the rapid expansion of gene-editing techniques, however, the difficulties of the procedure seem to rule out this option, making traceability irrelevant.

The debate has several consequences for access and benefit-sharing. These include whether DSI or *in silico* genetic resources fall within the scope of the biodiversity legal regime and the ABS mechanism established in the NP; whether the non-monetary and general scientific benefits arising from open access to DSI can be considered sufficient as an ABS mechanism; and, most crucially, whether an ABS regime is the best way to ensure that the use of DSI benefits provider countries, science and biodiversity conservation (Laird et al. 2020, p. 1200). There may be agreement that access to DSI for non-commercial research should be open, thus favouring scientific progress and open science, although open science is not always truly accessible to States that do not have the necessary capacities and it therefore does not necessarily mean fair science (Aubry 2019, p. 3). Moreover, in relation to commercial uses, there are even greater divergences on the sharing of benefits arising from the commercial use of the results of DSI utilization (FAO 2018b).

A similar debate is taking place within the framework of ITPGRFA reform, where the recent development of DSI has also had an impact (Aubry 2019). Indeed, the desirability of a new and reformed SMTA to incorporate a new definition of genetic parts and components that includes DSI in relation to plant genetic resources (FAO 2017) has been raised, but no agreement has been reached at present. Similarly, the FAO CGRFA is set to examine the issue of DSI at its next meeting in 2021 (FAO 2019b).

Finally, at CoP 13 in 2016, an information-gathering process on DSI was launched (CBD 2016b) with the creation of the *Ad Hoc* Technical Expert Group (AHTEG). At CoP 14 in 2018, the potential importance of DSI was recognized in relation to all three mutually supportive objectives of the Convention, although there was a call for greater conceptual clarity on DSI. However, given the existence of divergent views among the Parties over the sharing of benefits arising from the use of digital information on genetic resource sequences, the decision was taken to set in motion a renewed process, both political and scientific in nature, to facilitate debate on the matter, specifically with a view to strengthening the fulfilment of the third objective of the Convention (CBD 2018c). To this end, the decision expanded the remit of the AHTEG, which was, among other aspects, commissioned to undertake a study of how national measures address the sharing of benefits arising from the commercial and non-commercial use of DSI relating to genetic resources. As set out in the decision, the recommendations of the expanded Technical Expert Group will be examined by the Open-ended Intersessional Working Group on the Post-2020 Global Biodiversity Framework that will address DSI in this context (CBD 2018e).

As a basic trend, it can be argued that the debate on DSI should be redirected towards the fair and equitable distribution of scientific advances and progress for all human beings. From a human rights perspective, there is no doubt that scientific progress in the field of life sciences and modern biotechnology, like all scientific progress, must be conceived with an understanding that everyone has a right to enjoy and share in scientific advancement and its benefits. The spirit of this proclamation, applicable also to DSI, is still present both in the notion of CHM and in the ABS systems established internationally.

7.6 Access and Benefit-Sharing in Relation to Marine Genetic Resources in Areas Beyond the Jurisdiction of States

In recent years, an international regulatory process on marine biodiversity in areas beyond national jurisdiction (ABNJ) has been launched within the United Nations. Among the dimensions to be negotiated, the process also incorporates ABS with respect to marine genetic resources (MGR). After preparatory work that included an

Ad Hoc Open-ended Working Group (UNGA 2004) with the potential to define the issues to be addressed in the future negotiating package, the full-scale process began in 2012 at the United Nations Conference on Sustainable Development (UNCSD). The UNCSD decided that it would be desirable to regulate the sustainable use and conservation of this biological diversity not under the sovereignty of States. As indicated earlier, the CBD, the NP and ITPGRFA are governed by the principle that genetic resources and, for our purposes here, the various ABS systems involved, are regulated on the basis of the sovereignty of States over genetic resources. Regardless of the difficulties and economic costs of their exploitation, the main places where there is an abundance of MGR that are not under the jurisdiction of any State, but that may be subject to bioprospecting and make important contributions to scientific progress and to new industrial and commercial uses, can be found on the high seas and in the international seabed area.

In this regard, there is no specific regulation to ensure access and sustainable use and conservation, nor is there an ABS system to govern use. Part XI of UNCLOS states that the area in question and its resources are CHM, but the passage refers exclusively to mining resources and does not apply to MGR. With respect to mineral resources, the Agreement of 28 July 1994 relating to the implementation of Part XI of the United Nations Convention on the Law of the Sea of 10 December 1982 largely dilutes the impact of the CHM concept. Moreover, on the one hand, there are doctrinal positions that support the idea that CHM should also encompass all natural resources, including biological ones like microorganisms (Salamanca Aguado 2006) and, on the other, UNCLOS does contain references to the benefits that can be derived from marine scientific research in relation to both living and non-living resources. First, when it is carried out in the area in question, it shall be carried out not only for peaceful purposes, but also for the benefit of mankind as a whole; the International Seabed Authority is required to promote and encourage the conduct of marine scientific research in the area, and coordinates and disseminates the results of this research and analysis when available (article 143 UNCLOS). In this respect, the transfer of technology and scientific knowledge shall be promoted so that all States benefit therefrom (article 144.1.b UNCLOS). Second, UNCLOS similarly establishes in relation to scientific research in general that the dissemination of information and knowledge resulting from marine scientific research shall be actively promoted, especially to developing States (article 244.2 UNCLOS).

The recommendations of the Working Group (United Nations 2011) and the results of the 2012 UNCSD prompted the General Assembly to set up a preparatory committee (UNGA 2015b). In its final report (UN 2017), the preparatory committee laid out the issues where there were divergences, particularly those relating to the ABS regime that could be established in the treaty and that amounted to what Vázquez Gómez has characterized as one of the most controversial issues in the negotiations (Vázquez Gómez 2019, pp. 1, 23). In the end, the General Assembly (UNGA 2017) came to the view that an Intergovernmental Conference could be convened to draft an international legally binding instrument under the United Nations Convention on the Law of the Sea to address the conservation and sustainable use of marine biological diversity in areas beyond national jurisdiction. Thus, the objective now is to negotiate the conservation and sustainable use of marine biological diversity in ABNJ, together and as a whole, marine genetic resources, including questions on the sharing of benefits, measures such as area-based management tools, including marine protected areas, environmental impact assessments and capacity-building, and the transfer of marine technology (UNGA 2017). The Intergovernmental Conference has thus far held three sessions and the latest revised draft text submitted by the Presidency (UN 2019) is currently being circulated.

In any case, major differences still remain between States over many aspects of the revised text presented by the Presidency, especially in relation to MGR and, in as far as it is relevant here, the establishment of an ABS mechanism. An initial point of contention concerns general principles or approaches. In this respect, the first controversial question is whether the CHM principle and the principle of equity (Scovazzi 2020, p. 219), as well as the idea of an ABS regime are included among the principles. The ABS mechanism was introduced in the negotiations as a way to redistribute wealth and technology to countries that lack the capacity to explore and exploit MGR in ABNJ. The proposals submitted in 2020 by a number of States show the varying approaches to the issue taken by, on one hand, the European Union, its Member States and the United States, for example, and, on the other hand, the States belonging to the Core Latin American Group, which continues to insist on "the inclusion of the principle of common heritage of mankind as the guiding principle of the interpretation and application" of the future instrument (UN 2020).

A second point of contention relates to the problem of managing access to MGR when they are in an ABNJ and therefore a PIC mechanism is neither feasible nor can be envisaged. In this respect, the various options on the table include proposals for a mechanism of prior or *ex post* notification, which may seem reasonable, and for a permit and licence mechanism, which may be much more difficult to accept. In any case, ABS will require a well-defined institutional instrument that goes beyond the notion of a clearing-house mechanism (Tvedt and Jørem 2013). In addition, another question now under discussion is whether the future agreement will apply only to MGR that are accessed or collected *in situ* or it will apply as well to MGR accessed *ex situ* or even to MGR that are accessible *in silico*.

Lastly, the option of a multilateral ABS mechanism could perhaps be considered, since a bilateral system in relation to genetic resources that do not fall under the sovereignty of any State is by definition inoperative, given that there are no provider countries. Some stakeholders prefer a limited governance approach to access, favouring a pre- or post-collection notification of MGR *in situ* (Collins et al. 2020, p. 9). In any event, it is important to note the controversy over the benefits that should be derived from an ABS system in the future agreement and whether they should be only non-monetary, in the sense of sharing and collaborating in scientific research, technology transfer and capacity-building, or also monetary. Some authors suggest that it would be more appropriate in general to provide exclusively non-monetary benefits of a mandatory nature and to establish that monetary benefits could also voluntarily be considered if MGR are commercialized (Collins et al. 2020, p. 14). In any case, as the revised text reflects, the options under negotiation on this and the earlier issues remain wide open.

7.7 Conclusion

In conclusion, I would like to stress, first, the key importance of genetic resources in a context of conservation and sustainable development of biodiversity, particularly at a time when scientific advances have given rise to immense potential benefits and exceptional challenges. These benefits and challenges operate within a global framework of profound inequalities of development, which affect the lives of millions of people, both in terms of sustainable development more generally and of food security more specifically. The recognition of the full sovereignty of States over their natural resources must be combined with the common interest of humanity in these resources, their sustainability and development, and their systemic benefits for all humankind. While contradictions remain, there can be said to be a certain consensus around this trend based on solidarity and sustainable development.

The international legal regime on biodiversity addresses the specifically redistributive dimension with the establishment of an ABS system that aims to reconcile the sovereign rights of States, the access of users to genetic resources, and the fair and equitable sharing of benefits. However, the regime, which pivots on the NP, is strictly bilateral, that is, between the State where the resources are found and the State where the potential users of the resources are located, whether they are scientific users or users with economic and commercial interests. In any case, initial assessments indicate that the results today are not satisfactory in terms of the return of benefits for all or for provider States, which are also developing States. Nor are the results satisfactory in relation to ITPGRFA. Accordingly, while these two ABS systems are only a few years old, it appears reasonable to argue that they should be strengthened or reconsidered as a whole.

Added to all this is the crucial and truly disruptive element of DSI, which raises the possibility that access to genetic resources is not necessary and therefore puts the entire ABS system at risk. Even now, negotiations are underway on an agreement covering ABNJ, which are areas that are potentially rich in MGR. At present, there remain many difficulties to reaching an agreement and many issues where the positions are still very divergent, especially with regard to an ABS system or similar procedure that would allow for the benefits arising from the use of MGR in ABNJ to be passed on to all States.

In this vein, it should be noted in conclusion that the next Conference of the Parties to the CBD (CoP 15) and the meeting of the Parties to the NP (MoP 4), which have both been postponed until the second quarter of 2021, will have as their central theme the adoption of the Post-2020 Global Biodiversity Framework. One of the issues that must now be addressed in this Post-2020 Framework is the ABS mechanism provided for in the current international biodiversity regime, if we are to promote mutually supportive implementation between the NP and ITGRFA ABS systems, provide necessary responses to the challenge of DSI, and strengthen legally effective, politically acceptable, environmentally sustainable and economically satisfactory ABS mechanisms.

References

Aubry, S. (2019). The future of digital sequence information for plant genetic resources for food and agriculture. *Frontiers in Plant Science*. Retrieved September 30, 2020, from https://www.frontiersin.org/articles/10.3389/fpls.2019.01046/full

Bouza, N., García, C., & Rodrigo, A. J., (dirs.). (2015). In *La gobernanza del interés público global*. Madrid: Tecnos.

Broggiato, A., Dedeurwaerdere, T., Batur, F., & Coolsaet, B. (2015). Introduction. Access benefit-sharing and the Nagoya Protocol: The confluence of Abiding Legal Doctrines. In B. Coolsaet et al. (Eds.), *Implementing the Nagoya Protocol: Comparing access and benefit-sharing regimes in Europe*. Leiden/Boston: Brill Nijhoff.

CBD. (2002). *Access and benefit-sharing as related to genetic resources. A. Bonn Guidelines on Access to Genetic Resources and Fair and Equitable Sharing of the Benefits Arising out of their Utilization*, Decision VI/24 A COP. Retrieved September 30, 2020, from https://www.cbd.int/decisions/?id=7198

CBD. (2016a). *Review of progress towards Aichi Biodiversity Target 16 on the Nagoya Protocol*, Decision 2/1 COP-MOP. Retrieved September 30, 2020, from https://www.cbd.int/doc/decisions/np-mop-02/np-mop-02-dec-01-en.pdf

CBD. (2016b). *Digital sequence information on genetic resources*, Decision 2/14 COP-MOP. Retrieved September 30, 2020, from https://www.cbd.int/doc/decisions/np-mop-02/np-mop-02-dec-14-en.pdf

CBD. (2016c). *Progress in the implementation of the Convention and the Strategic Plan for Biodiversity 2011–2020 and towards the achievement of the Aichi Biodiversity Targets*, Decision XIII/1 COP. Retrieved September 30, 2020, from https://www.cbd.int/doc/decisions/cop-13/cop-13-dec-01-en.doc

CBD. (2018a). *Assessment and review of the effectiveness of the Protocol (Article 31)*, Decision 3/1 COP-MOP. Retrieved September 30, 2020, from https://www.cbd.int/doc/decisions/np-mop-03/np-mop-03-dec-01-en.pdf

CBD. (2018b). *Global multilateral benefit-sharing mechanism (Article 10)*, Decision 3/13 COP-MOP. Retrieved September 30, 2020, from https://www.cbd.int/doc/decisions/np-mop-03/np-mop-03-dec-13-en.pdf

CBD. (2018c). *Digital sequence information on genetic resources*, Decision 14/20 COP. Retrieved September 30, 2020, from https://www.cbd.int/doc/decisions/cop-14/cop-14-dec-20-en.docx

CBD. (2018d). *A Fact-Finding and Scoping Study on Digital Sequence Information on Genetic Resources in the Context of the CBD and the Nagoya Protocol*, S.A. Laird and R.P. Wynberg, with contributions from A. Iranzadeh and A. Sliva Kooser, CBD/DSI/AHTEG/2018/1/3.

CBD. (2018e). *Digital sequence information on genetic resources Digital sequence information on genetic resources*, Decision 3/12 COP-MOP. Retrieved September 30, 2020, from http://www.cbd.int/doc/decisions/np-mop-03/np-mop-03-dec-12-en.pdf

CBD. (2019). *Including Access and Benefit-sharing in the Post-2020 Global Biodiversity Framework*, CBD/SBSTTA/23/INF/10.

CGIAR et al. (2020). *Including Access and Benefit Sharing in the Post-2020 Global Biodiversity Framework*. Retrieved September 30, 2020, from https://cgspace.cgiar.org/bitstream/handle/10568/106248/http://ABI-1.pdf

Collins, J. E., Vanagh, Th., & Huys, I. (2020). Stakeholder perspectives on access and benefit-sharing for areas beyond national jurisdiction. *Frontiers in Marine Science*. Retrieved September 30, 2020, from https://www.frontiersin.org/articles/10.3389/fmars.2020.00265/full

EJIL. (2012). *Symposium: Global Public Goods and the Plurality of Legal Orders*, Issue Vol. 23 (2012) 3, 643–791.

FAO. (1983). *International Undertaking on Plant Genetic Resources*, Resolution 8/83 FAO Conference.

FAO. (2001). *Adoption of the International Treaty on Plant Genetic Resources for Food and Agriculture and Interim Arrangements for its Implementation*, Resolution 3/2001 FAO Conference.

FAO. (2013). *Report of the Governing Body of the International Treaty on Plant Genetic Resources for Food and Agriculture*, IT/GB-5/13/Report.

FAO. (2017). *Report of the Governing Body of the International Treaty on Plant Genetic Resources for Food and Agriculture*, IT/GB-7/17/Report, Resolution 13/2017.

FAO. (2018a). *Exploratory Fact-Finding Scoping Study on 'Digital Sequence Information' on Genetic Resources for Food and Agriculture*, by J.A. Heinemann, D.S. Cray and D.S. Thaler, Background Study paper No. 68.

FAO. (2018b). *Report of the Fourth Session of the Team of Technical and Legal Experts on Access and Benefit-sharing*, CGRFA/TTLE-ABS-4/18/Report.

FAO. (2019a). *Report of the Governing Body of the International Treaty on Plant Genetic Resources for Food and Agriculture*, IT/GB-8/19/Report, Resolution 4/2019.

FAO. (2019b). *Report of the Commission on Genetic Resources for Food and Agriculture*, Appendix B, Elements to Facilitate Domestic Implementation of Access and Benefit-sharing for Different Subsectors of Genetic Resources for Food and Agriculture, with explanatory notes (ABS Elements), CGRFA-17/19/Report.

Karger, E., Plessis, P., & Meyer, H. (2019). *Digital sequence information on genetic resources*. Eschborn: ABS Initiative.

Kate, K., & Laid, S. A. (2000). *The commercial use of biodiversity. Access to genetic resources and benefit-sharing*. London: Earthscan.

Laird, S., Wynberg, R., Rourke, M., Humphries, F., Ruiz Muller, M., & Lawson, C. (2020). Rethink the expansion of access and benefit sharing. *Science, 367*(6483), 1200–1202.

Morgera, E. (2016). The need for an international legal concept of fair and equitable benefit sharing. *European Journal of International Law, 27*(2), 353–383.

Morgera, E. (2018). Fair and equitable benefit-sharing. In L. Krämer & E. Orlando (Eds.), *Principles of environmental law* (pp. 323–337). Cheltenham: Ewdard Elgar Publishing.

Morgera, E., Buck, M., & Tsiounami, E. (2010). *The 2010 Nagoya Protocol on access and benefit-sharing in perspective: Implications for international law and implementation challenges*. Leiden: Martinus Nijhoff Publishers.

Morgera, E., Tsiounami, E., & Buck, M. (2014). *Unravelling the Nagoya Protocol: Commentary on the protocol on access and benefit-sharing to the CBD*. Leiden/Boston: Brill.

Pérez Salom, J. R. (2002). *Recursos Genéticos, Biotecnología y Derecho Internacional. La distribución justa y equitativa de beneficios en el Convenio sobre Biodiversidad*. Cizur Menor: Aranzadi/Thomson.

Salamanca Aguado, E. (2006). El régimen jurídico aplicable a la bioprospección de los recursos genéticos de los fondos marinos fuera de los límites de la jurisdicción nacional. *Revista Española de Derecho Internacional, 52*(2), 1072–1078.

Scovazzi, T. (2020). The rights to genetic resources beyond national jurisdiction: Challenges for the ongoing negotiations at the UN. In C. Barret (Ed.), *The law of the seabed: Access, users and protection of seabed resources* (pp. 213–237). Leiden: Brill/Nijhoff.

Sirakaya, A. (2019). Balanced options to access and benefit-sharing: Stakeholder insights on provider country legislation. *Frontiers in Plant Science*. Retrieved September 30, 2020, from https://doi.org/10.3389/fpls.2019.01175

Tvedt, M. W., & Jørem, A. E. (2013). Bioprospecting in the high seas: Regulatory options for benefit sharing. *The Journal of World Intellectual Property, 16*(3–4), 150–167.

UNGA. (1962). *Permanent sovereignty over natural resources*, Resolution 1803 (XVII), of 14 December 1962, of General Assembly.

UNGA. (1970). *Declaration of Principles Governing the Seabed and the Ocean Floor, and the Subsoil Thereof, beyond the Limits of National Jurisdiction*, Resolution 2749 (XXV), of 17 December 1970, of General Assembly.

UNGA. (1974a). *Declaration on the Establishment of a New International Economic Order* and *Programme of Action on the Establishment of a New International Economic Order*, Resolutions 3201 (S-VI) and 3202 (S-VI), of 1 May 1974, of General Assembly.
UNGA. (1974b). *Charter of Economic Rights and Duties of States*, Resolution 3281 (XXIX), of 12 December 1974, of General Assembly.
UNGA. (1948). Universal Declaration of Human Rights, Resolution 217 (III) A, of 10 December 1948, of General Assembly.
UNGA. (1986). *Declaration on the right to development*, Resolution 41/128, of 4 December 1986, of General Assembly.
UNGA. (2004). *Oceans and the law of the sea*, Resolution 59/24, of 17 November 2004, of General Assembly.
UNGA. (2007). United Nations Declaration on the Rights of Indigenous Peoples, Resolution 61/295, of 13 September 2007, of General Assembly.
UNGA. (2015a). *Transforming our world: the 2030 Agenda for Sustainable Development*, Resolution 70/1, of 25 September 2015, of General Assembly.
UNGA. (2015b). *Development of an international legally binding instrument under the United Nations Convention on the Law of the Sea on the conservation and sustainable use of marine biological diversity of areas beyond national jurisdiction*, Resolution 69/292, of 19 June 2015, of General Assembly.
UNGA. (2017). *International legally binding instrument under the United Nations Convention on the Law of the Sea on the conservation and sustainable use of marine biological diversity of areas beyond national jurisdiction*, Resolution 72/249, of 24 December 2017, of General Assembly.
United Nations. (1987). *Report of the World Commission on Environment and Development: Our Common Future*, A/42/427.
United Nations. (1992). *Report of the United Nations Conference on Environment and Development*, A/CONF.151/26/Rev.1 (Vol. I).
United Nations. (2002). *Report of the World Summit on Sustainable Development*, A/CONF.199/20.
United Nations. (2011). *Letter dated 30 June 2011 from the Co-Chairs of the Ad Hoc Open-ended Informal Working Group to the President of the General Assembly*, A/66/119.
United Nations. (2017). *Report of the Preparatory Committee established by General Assembly resolution 69/292: Development of an international legally binding instrument under the United Nations Convention on the Law of the Sea on the conservation and sustainable use of marine biological diversity of areas beyond national jurisdiction*, A/AC.287/2017/PC.4/2.
United Nations. (2019). *Revised draft text of an agreement under the United Nations Convention on the Law of the Sea on the conservation and sustainable use of marine biological diversity of areas beyond national jurisdiction*, A/CONF.232/2020/3.
United Nations. (2020). *Textual proposals submitted by delegations by 20 February 2020, for consideration at the fourth session of the Intergovernmental conference on an international legally binding instrument under the United Nations Convention on the Law of the Sea on the conservation and sustainable use of marine biological diversity of areas beyond national jurisdiction (the Conference), in response to the invitation by the President of the Conference in her Note of 18 November 2019*. Retrieved September 30, 2020, from https://www.un.org/bbnj/sites/www.un.org.bbnj/files/textual_proposals_compilation_-_15_april_2020.pdf
Vázquez Gómez, E. (2019). La protección de la diversidad biológica marina más allá de la jurisdicción nacional. Hacia un nuevo acuerdo de aplicación de la Convención de las Naciones Unidas sobre el Derecho del Mar. *Revista Electrónica de Estudios Internacionales*. 37. http://www.reei.org/index.php/revista/num37/articulos/proteccion-diversidad-biologica-marina-mas-alla-jurisdiccion-nacional-hacia-nuevo-acuerdo-aplicacion-convencion-naciones-unidas-sobre-derecho-mar

Part III
Some Challenges for Sustainability and Marine Biodiversity

Chapter 8
Sub-National Island Jurisdictions and Biodiversity Protection: Moving Towards Greater Resilience and Sustainability

Francesco Sindico

Contents

8.1 Introduction 141
8.2 Sub-National Island Jurisdictions and the Convention on Biological Diversity 143
8.2.1 Sub-National Island Jurisdictions and the Åland Islands 143
8.2.2 Island Biodiversity and the Convention on Biological Diversity 144
8.2.3 Sub-National Island Jurisdictions and the Convention on Biological Diversity . 146
8.3 The Åland Islands, Biodiversity and Sustainable Development 149
8.4 Island Resilience and Sustainability and Biodiversity Protection 150
8.5 Conclusion 152
References 154

Abstract This chapter discusses the extent to which sub-national island jurisdictions (SNIJs) can rely on the UN Biodiversity Convention to become more resilient and sustainable. The chapter focuses on the Åland Islands in Finland as an example of an SNIJ that has developed a sustainability agenda that promotes resilience and includes biodiversity protection. The Åland Islands provide an example of how international legal processes and agreed objectives, such as the SDGs and to a lesser extent the CBD, can be incorporated into and drive forward resilience and sustainability agendas that include biodiversity protection and take bottom-up approaches.

I would like to thank Nicola Sharman for her research assistance in the preparation of the chapter.

F. Sindico (✉)
University of Strathclyde, Law School, Glasgow, UK
e-mail: francesco.sindico@strath.ac.uk

M. Campins Eritja, T. Fajardo del Castillo (eds.), *Biological Diversity and International Law*, https://doi.org/10.1007/978-3-030-72961-5_8

8.1 Introduction

Biodiversity can be regarded as "the variability among living organisms from all sources including, inter alia, terrestrial, marine and other aquatic ecosystems and the ecological complexes of which they are part; this includes diversity within species, between species and of ecosystems" (CBD 1992, art. 2). There are three reasons to promote island biodiversity protection. First, biodiversity should be protected *per se*. Second, a healthy biodiversity is a requirement for a strong economy and good quality of life. That is, biodiversity protection needs to be given the utmost importance by policy-makers and decision-makers because of the services that a rich biodiversity gives to island communities, ranging from healthy fish stocks to vibrant land-based ecosystems, to name but a few (CBD 2006, para. 20). Third, protecting island biodiversity will not only bring in more tourists, but can also foster a different, more sustainable kind of tourism.

Machair offers one example of the need to protect biodiversity on all three grounds. While machair is difficult to define, it is clearly more than just a habitat. Rather, it is a patch of land that presents a unique combination of flora and fauna found only in the Western Isles of Scotland and some parts of Ireland (Nature Scot 2020). Machair is important for the environment because it sustains several scarce species. At the same time, machair provides fertile land to crofters who benefit from working the land. Moreover, a healthy machair attracts tourists because of the beautiful colours of its flowers when they come into bloom. Machair is by no means unique, but it nicely illustrates the reasons for protecting biodiversity on islands and the benefits that can ensue.

Given this background, the chapter explores whether sub-national island jurisdictions (SNIJs) can rely on the UN Convention on Biological Diversity (CBD) in their efforts to develop and implement development agendas that will ultimately make their islands more resilient and sustainable.

The chapter is divided into five sections. After the introduction, the second section introduces the category of SNIJs, discusses the extent to which the CBD provides for island biodiversity within its regime, and enquires into whether the CBD recognizes SNIJs as meaningful biodiversity actors. The third section presents the Åland Islands as an example of a European SNIJ, highlights their efforts to promote sustainable development, and looks at the extent to which biodiversity protection is part of such efforts. The fourth section discusses the concepts of resilience and sustainability within current island studies in the context of biodiversity protection. The final objective of the section is to determine the extent to which SNIJs can use the CBD to make their islands more resilient and more sustainable. Lastly, the chapter draws some general conclusions and highlights areas for future research.

8.2 Sub-National Island Jurisdictions and the Convention on Biological Diversity

This section opens with a discussion of the category of SNIJs, explaining that the Åland Islands are indeed an SNIJ. The rest of the section is devoted to an analysis of the CBD to determine whether island biodiversity is part of the regime, followed by an exploration of the extent to which SNIJs can use the CBD in their efforts to promote biodiversity protection.

8.2.1 Sub-National Island Jurisdictions and the Åland Islands

To understand what SNIJs are, we need to determine the meaning of two words: "sub-national" and "jurisdictions". By sub-national, we refer to islands that do not constitute a state by themselves. Hence, Small Island Developing States, by virtue of being states, are not SNIJs. The word "jurisdictions" implies that the islands possess some degree of normative power over decisions that affect their livelihood (Warrington and Milne 2018, pp. 174–177). Such islands could be "states" within a federal country, such as Hawai'i in the United States, or a province, such as Prince Edward Island in Canada, or a region in a non-federal state, such as Sicily in Italy. More importantly, an SNIJ will have some degree of local representation that permits its island community to take, or at least inform, decisions adopted centrally, as is the case of the Orkney and Shetland Islands in Scotland (Depraetere and Dahl 2018, map 2.11).

The focus of this chapter is the Åland Islands. The Åland Islands are an archipelago of 6700 islands in the Baltic Sea that belong to Finland. The archipelago is an autonomous Swedish-speaking region of Finland, with its own parliament and government (Lindstrom 2000). It is also a demilitarised territory (Sindico 2017, p. 1). Warrington and Milne (2018, p. 177) refer to the Åland Islands as an example of islands "represented in special federal relationships" akin to the Faroe Islands and the Isle of Man. When acceding to the EU as part of Finland, the Åland Islands were able to secure special treatment "permitting duty-free transactions on Åland ships within EU waters and confirming restrictions on the right of non-residents to own property" (Warrington and Milne 2018, p. 179). The Åland Islands have many other characteristics related to their normative constitutional status and their socio-economic patterns, but the short remarks above suffice to frame them as an SNIJ.

Shortly, the chapter will return to the Åland Islands in order to discuss more specifically their biodiversity and the extent to which biodiversity protection is part of their sustainable development agendas. First, however, it is necessary to examine the extent to which the CBD devotes special attention to island biodiversity more broadly.

8.2.2 *Island Biodiversity and the Convention on Biological Diversity*

The ultimate objective, main provisions, institutional framework and legal nature of the CBD are dealt with elsewhere in this book[1] (Secretariat of the Convention on Biological Diversity 2005). The focus in this subsection is on the extent to which the CBD devotes attention to islands and then, in the next subsection, on the extent to which SNIJs could potentially rely on the CBD.

Contrary to other international treaties (UNFCCC 1994, preamble and Art. 4.8(a); Paris Agreement 2015, Art. 4.6, 9.4, 9.9, 11.1 and 13.3), the CBD makes no specific reference to islands in its provisions. However, in 2006, a CoP Decision was adopted on island biodiversity and the related programme of work sought to significantly reduce

> island biodiversity loss by 2010 and beyond at global, regional and national levels, through the implementation of the three main objectives of the Convention, for the benefit of all forms of life on islands and, in particular, as a contribution to poverty alleviation and the sustainable development of small island developing States (CBD 2006, para. 14).

The programme of work included almost 50 island-specific priority actions arranged under 11 goals (CBD 2006, Annex, Part E),[2] which in turn were divided into the following seven focal areas:

1. Protect the components of biodiversity
2. Promote sustainable use
3. Address threats to biodiversity
4. Maintain goods and services from biodiversity to support human well-being
5. Protect traditional knowledge and practices
6. Ensure the fair and equitable sharing of benefits arising out of the use of genetic resources
7. Ensure provision of adequate resources

Three key aspects of the CBD island biodiversity programme of work deserve attention: its encouragement of Parties to try to better understand biodiversity, its call for coordination with other international agreements, and its emphasis on the role of legal frameworks.

---

[1]See Chap. 2.

[2]The 11 goals are: (1) promote the conservation of the biological diversity of island ecosystems, habitats and biomes; (2) promote the conservation of island species diversity; (3) promote the conservation of island genetic diversity; (4) promote sustainable use and consumption; (5) pressures from habitat loss, land-use change and degradation, and sustainable water use, reduced on islands; (6) control threats to island biological diversity from invasive alien species; (7) address challenges to island biodiversity from climate change, and pollution; (8) maintain capacity of island ecosystems to deliver goods and services and support livelihoods; (9) maintain socio-cultural diversity of indigenous and local communities on islands; (10) ensure the fair and equitable sharing of benefits arising out of island genetic resources; and (11) parties have improved financial, human, scientific, technical and technological capacity to implement the Convention.

First, the CBD island biodiversity programme of work encouraged Parties to better understand the role of biodiversity within an island context and the drivers leading to its deterioration. The programme of work acknowledged the pressures stemming from activities such as tourism, agriculture, forestry and fisheries (CBD 2006, Appendix, para. 4.1.1.4). At the same time, it also highlighted the value that biodiversity has for livelihoods, the economy and the culture of people living on islands (CBD 2006, Appendix, para. 4.1.1.5). As mentioned above, priority actions were developed in order to meet the goals set out in the programme of work. Many of the priority actions focused on areas identified as putting pressure on island biodiversity. As a consequence, these areas stopped being defined as challenges and instead became opportunities for islands and their communities to adopt sustainable and integrated plans in fields such as marine resources, agriculture, forestry, tourism, to name but a few.[3]

Second, the programme of work made clear that, for the actions to be effective, coordination was necessary between the CBD and other international legal regimes. For example, support for island protected areas could be leveraged through designations provided for in the Ramsar and the World Heritage Conventions (CBD 2006, Appendix, para. 1.2.1.3.b). More effective protection of endemic species on islands could be achieved through synergies with the Convention on Migratory Species (CBD 2006, Appendix, para. 1.2.1.7). Furthermore, the development of transboundary marine protected areas should be conducted in line with guidance set out in the law of the sea (CBD 2006, Appendix, para. 1.2.1.8).

Third, the programme of work put emphasis on the role of law and legislative and policy frameworks. Legislation and management plans were to be put in place in a participatory manner with true input from a wide range of island stakeholders (CBD 2006, Appendix, para. 1.1.1.2). Only through the participation of local communities would there be a pull towards societal compliance with the laws put in place (CBD 2006, Appendix, para. 1.1.1.4). Moreover, the programme of work called for the removal of any negative incentives (i.e. subsidies) whose effect was to promote unsustainable socio-economic patterns that had an adverse impact on an island's biodiversity (CBD 2006, Appendix, para. 4.1.1.1).

In 2012, a further CoP Decision was adopted to review the island biodiversity programme of work (CBD 2012). The review was implemented by the Global Island Partnership (CBD 2008, para. 3). Since 2012, the CoP has taken no further action in the specific domain of island biodiversity. However, the 2012 CoP Decision did

[3]Other priority actions include the following: priority action 4.2.1. marine based resources; priority action 4.2.2. sustainably use terrestrial and freshwater biological diversity; priority action 4.2.3. sustainable and integrated agriculture development plan; priority action 4.2.4. sustainable forestry plan; priority action 4.2.5. mainstream biodiversity into the integrated planning, strategies, policies and implementation for all tourism and tourism-related projects; priority action 5.1.1. integrated land and water use plans; priority action 6.1.1. protect against the introduction of invasive alien species taking into account existing control systems, such as those under the International Plant Protection Convention; priority action 7.2.1. enforce the environmental, social and cultural impact assessment process for island industries, infrastructure and urban plans.

highlight the importance of the island biodiversity programme of work for the fulfilment of the Aichi Biodiversity Targets[4] and gave specific mention to Target 12, which focuses on the prevention of species extinction and their conservation (CBD 2012, preamble). The CoP Decision also reaffirmed the importance of legal frameworks to address biodiversity protection on islands (CBD 2012, preamble).

It is a fact that the international community has failed to fully achieve most of the Aichi Biodiversity Targets (Schroder 2020; Secretariat of the Convention on Biological Diversity 2020). The focus now is on the adoption of a post-2020 Global Biodiversity Framework at CoP 15 in 2021[5] (CBD 2018a), which should build on the 2050 Vision of "Living in harmony with nature", according to which:

> [B]y 2050, biodiversity is valued, conserved, restored and wisely used, maintaining ecosystem services, sustaining a healthy planet and delivering benefits essential for all people (CBD 2018b, para. 6).

8.2.3 Sub-National Island Jurisdictions and the Convention on Biological Diversity

Despite the fact that momentum on island biodiversity, as a specific programme of work within the CBD, appears to have slowed in recent years, the question remains as to whether SNIJs can still rely on the CBD in the furtherance of their biodiversity policies. By relying on the CBD, I mean the extent to which an SNIJ is encouraged and authorised, if not mandated, by the CBD to take proactive steps to protect its biodiversity.

The CBD, like many other areas of international environmental law, has acknowledged the relevance of sub-national actors and non-state actors (e.g. UNFCCC 2011, para. 7; UNFCCC 2015, para. 120; Hale 2018; Nolan 2016). States do not have the capacity or the resources to deal with biodiversity protection on their own. Consequently, there is an urgent need for engagement at all scales of governance to deal with the loss of biodiversity. The island biodiversity programme of work recognised this reality within islands by declaring to:

> [s]upport subnational implementation of the Convention in islands by engaging subnational and local authorities through the Plan of Action for Subnational Governments, Cities and Other Local Authorities for Biodiversity, adopted in decision X/22, and as informed by the

---

[4]The Aichi Biodiversity Targets have five strategic goals: Strategic Goal A: Address the underlying causes of biodiversity loss by mainstreaming biodiversity across government and society; Strategic Goal B: Reduce the direct pressures on biodiversity and promote sustainable use; Strategic Goal C: Improve the status of biodiversity by safeguarding ecosystems, species and genetic diversity; Strategic Goal D: Enhance the benefits to all from biodiversity and ecosystem services; and Strategic Goal E: Enhance implementation through participatory planning, knowledge management and capacity building.

[5]See Chap. 2.

> assessment of the links and opportunities between urbanization and biodiversity (the "Cities and Biodiversity Outlook") (CBD 2012, para. 3c).

Island-based "subnational and local authorities" can be SNIJs, which are called to implement the CBD in their own island context. The Plan of Action on Sub-national Governments, Cities and Other Local Authorities for Biodiversity (2011–2020) (CBD 2010) becomes crucial in this respect and warrants specific attention.

The overall goal of the Plan of Action is to foster increased cooperation between central governments and sub-national and local authorities in the implementation of the Convention, CoP Decisions and Aichi Biodiversity Targets (CBD 2010, Annex, para. 1). In particular, CoP Decision X/22 maintains the following:

> Parties to the Convention on Biological Diversity should, as appropriate, seek to engage their subnational Governments, cities and other local authorities, as appropriate, to achieve the objectives of the Convention and the implementation of the Strategic Plan for Biodiversity 2011–2020, by developing policy tools, guidelines and programmes, providing technical assistance and/or guidance, as appropriate, in line with their national biodiversity strategies and action plans (NBSAPs) and other relevant governance arrangements established by their national Governments (CBD 2010, Annex, para 2).

Four key observations on the objective and operationalisation of the Plan of Action are relevant from an SNIJ perspective: the need for SNIJs to have the necessary constitutional powers to deliver on the Plan of Action; a bottom-up approach with effective public participation; SNIJs as hubs of innovation that are capable of sharing good practices; and the integration of biodiversity into wider sustainability plans.[6]

First, coordinating between central government and the periphery in order to deliver effective biodiversity protection requires an SNIJ to have the necessary governance structures. It is not enough for the CBD to encourage sub-national governments and local authorities to play a role in the implementation of biodiversity plans and strategies. Sub-national governments and local authorities need to

[6]The CBD (2010, Annex, para. 3) gives more detail on the mission provided for in the Plan of Action. Specifically, by 2020: (a) relevant tools, guidelines and capacity-building programmes based on best practices, as well as innovative financial mechanisms to support their implementation should, as appropriate, be in place to increase synergies between the various levels of government in implementing the Strategic Plan for Biodiversity 2011–2020, considering the specific mandates of each level of government; (b) national biodiversity strategies and action plans should be supported, as appropriate, by sub-national and local strategies and corresponding action plans; (c) awareness campaigns on the importance of biodiversity and ecosystem services should, as appropriate, be implemented at local level as part of the Parties' communication, education and public awareness strategies, including major groups such as business, youth, non-governmental organizations and indigenous and local communities, through initiatives such as celebrations of the International Day for Biological Diversity (May 22), the Green Wave initiative, and other activities in support of the Convention on Biological Diversity; and (d) monitoring and evaluation systems for sub-national governments and local authorities should, as appropriate, be applied, guided by national frameworks, to report on progress to national governments in line with reporting obligations under the Convention on Biological Diversity, and to set benchmarks for local biodiversity management in line with the 2011–2020 indicator framework under the Convention on Biological Diversity, using tools such as the Singapore Index on Cities' Biodiversity.

have proper competences from central government in order to be able to deliver such implementation.

Second, the possibility envisaged in the Plan of Action for sub-national governments and local authorities to play a role in implementing the CBD and related decisions and targets is a much-needed acknowledgment that local communities will often know better than central government how to deal with challenges and opportunities arising from the biodiversity in their land and waters. This bottom-up approach, whereby locals know more than remote officials, is a promising policy avenue, but only if local communities are given a proper voice. This is where consultation and effective public participation come into play (Sindico and Crook 2021). All good efforts to build bottom-up biodiversity policies in SNIJs will come tumbling down like a house of cards, not only if the sub-national government or local authority lacks the necessary competences, but also if public participation ends up being a tick-box exercise without effective input into the strategies and policies that deal with biodiversity in the island context.

Third, it is not only SNIJs and the people living on the islands who benefit from the opportunity given by the CBD for them to get involved in the implementation of the CBD and related decisions and targets. Global biodiversity discourse, too, gains from the sharing of lessons and good practices by sub-national governments and local authorities. In fact, islands are often seen as test-beds for innovation and the ways in which SNIJs deal with biodiversity loss are no exception (CBD 2010, Annex, para. 4.b).

Fourth, the Plan of Action clarifies that, in order to deal efficiently with the loss of biodiversity, an integrated approach needs to be taken so that biodiversity protection is included in wider sustainability plans (CBD 2010, Annex, para. 5.e). This is particularly important and true within island settings, where life is sometimes even more interconnected than in mainland settings. A further aspect of integration is the need to explore synergies and coordination between the CBD and other international legal instruments that cover biodiversity protection and allow for the participation of sub-national governments and local authorities (CBD 2010, Annex, para. 5.c).

In this vein, the next section highlights how the four key aspects of the Plan of Action on Sub-national Governments, Cities and Other Local Authorities for Biodiversity are present in the Åland Islands. The Åland Islands, which have the necessary constitutional powers to take a proactive role, have taken a bottom-up approach with effective public participation to develop their sustainability agenda, which includes biodiversity protection. Indeed, the Åland Islands can be framed as a hub of innovation whose good practices are worthy of sharing. Finally, biodiversity is not considered as a stand-alone policy area in the Åland Islands, but rather as part of an integrated approach to sustainable development.

8.3 The Åland Islands, Biodiversity and Sustainable Development

One of the main challenges to biodiversity on the Åland Islands comes from unsustainable planning practices that have led to the construction of houses either directly on the shore or very nearby. A long-lasting tension between private right holders and the protection of the environment lies at the origin of these practices. The Åland Islands also have a rather poor performance when it comes to the percentage of protected land area, which was only 2.7% in 2017 compared with the European average of between 10% and 15% (Bärkraft 2017, p. 22).

Against this backdrop, both the parliament and the government of the Åland Islands adopted a landmark decision in 2014 aimed at achieving total sustainable development by 2051. A bottom-up, community-led process resulted in the adoption, in 2016, of a development and sustainability agenda (Bärkraft 2016) that includes seven strategic goals to achieve by 2030:

1. Happy people whose inherent resources increase
2. Everyone feels trust and has real possibilities to participate in society
3. All water is of good quality
4. Ecosystems in balance and biological diversity
5. Attractive for residents, visitors and businesses
6. Significantly higher proportion of energy from renewable sources, plus increased energy efficiency
7. Sustainable and mindful patterns of consumption

The development and sustainability agenda is mindful of the Sustainable Development Goals (SDGs) and the parallel timeline between the SDGs and the process initiated in the Åland islands. In fact, the local development and sustainability agenda clarifies that each of its seven goals contributes to meeting one or more of the SDGs.

The development and sustainability agenda also has a specific biodiversity goal, goal 4, which calls for ecosystems to be "protected and used in a sustainable way by 2030". This shows a clear recognition of the strong link between healthy biodiversity and resilience. In fact, the agenda explicitly states that "biological diversity contributes to increased resilience in ecosystems" (Bärkraft 2016, p. 14).

In addition, the agenda touches on many of the pressures on biodiversity highlighted by the CBD island biodiversity programme of work and recognises some of the points highlighted in the Plan of Action on Sub-national Governments, Cities and Other Local Authorities for Biodiversity (2011–2020) (CBD 2010). The integration of biodiversity in development and planning strategies is seen as a key tool to achieve greater protection of ecosystems. The agenda also recognises the important role of agriculture, forestry and fisheries in protecting island biodiversity.

Despite the inclusion of the areas and activities mentioned above, the development and sustainability agenda of the Åland Islands makes no explicit reference to the CBD. However, the CBD does receive mention in the 2017 status report as an instrument that the Åland Islands must follow (Bärkraft 2017, p. 23). The opposite can be said of the SDGs, which appear to have been expressly used to drive the process and have served, in some cases, as possible benchmarks, although the goals and indicators in the agenda have then been localized to the Åland Islands context. Strategic development goal 4 refers specifically to SDG 14 in stating that by 2030 "at least 10% of coastal and marine environments are protected" (Bärkraft 2016, p. 14).

The Åland Islands development and sustainability agenda provides an example of how an SNIJ can effectively engage in the protection of biodiversity within its own island/s, provided that a number of conditions are in place. These conditions were highlighted in the Plan of Action on Sub-national Governments, Cities and Other Local Authorities for Biodiversity (2011–2020): possession of the necessary legal competences, effective public participation and integration.

First, the constitutional landscape of Finland provides the Åland Islands with the necessary autonomy and competences to take an active stand in relation to local biodiversity. Second, the development and sustainability agenda is a prime example of policy delivered through an effective bottom-up approach. In fact, the use of a bottom-up process to develop such an agenda is a recognised good practice (Bärkraft 2019). By relying on robust public participation from the outset, the agenda is "owned" by the local community, making it likelier that implementation will lead to more effective outcomes. Third, the specific biodiversity goal in the agenda cannot be met in isolation. Integration is key to success and, just as with the SDGs themselves, several, if not all, of the goals set out in the 2030 development and sustainability agenda are interlinked and dependant on one another. For example, good water quality (goal 3) is paramount for ecosystems (goal 4) and necessary to attract visitors (goal 5).

Overall, the presence of the SDGs—and, to a more limited extent, the CBD—in an official policy of a sub-national government like the Åland Islands is a testament to how SNIJs can rely on the CBD and the SDGs to promote biodiversity protection. However, to what extent does the interaction between international legal instruments, such as the CBD and the SDGs, and SNIJs promote biodiversity protection in such a way that it also encourages island resilience and sustainability? The next section discusses this question in greater detail.

8.4 Island Resilience and Sustainability and Biodiversity Protection

In this final substantive section, the focus turns to how resilience and sustainability are framed within island studies and how they relate to biodiversity protection.

According to an ecology approach to resilience (Begon et al. 2003), there can be "two nuanced meanings of resilience: (1) how quickly a system might return to stability after being disturbed; and (2) the extent to which a system can be disturbed without breaking down" (Kelman and Randall 2018, p. 354). Such an approach to resilience, however, has been criticized within the island studies literature as not sitting comfortably with island realities, since such realities are not static (Kelman and Randall 2018, p. 354). Quite the opposite, islands are dynamic places where being resilient often means having the capacity to change. Resilience should be considered not so much as the capacity of an island and its community to return to an original state after some disturbance by a natural or human shock. Rather, a resilient island will be an island that not only is able to bounce back from a crisis, but can also do so in a way that promotes a thriving society (Kelman and Randall 2018, p. 360). If considered in this different manner, resilience starts to close the gap with sustainability. In fact, if resilience is about driving an agenda for a better island in the wake of a shock, then the question becomes what kind of future does the island want?

A sustainable island is an option that will be amenable to island governments and communities depending on what is meant by sustainable. Since the late eighties, however, sustainability has become a notion that is highly politically charged. From one of the first definitions of sustainable development that appeared in the Brundtland Report in 1987 (World Commission on Environment and Development 1987) to later iterations in the United Nations Conference on Environment and Development in 1992 (Rio Declaration 1992, Principles 1 and 3), in the Johannesburg Summit on Sustainable Development in 2002 (Johannesburg Declaration on Sustainable Development 2002) and finally in the 2030 Agenda adopted in 2015 by the UN General Assembly (UNGA 2015), it is now clear that three areas need to be considered for a future to be sustainable: economic, environmental and social. In other words, a business-as-usual societal model that focuses only on economic growth while paying little attention to the effects on the environment and retaining, or worse, increasing social inequality is not sustainable. Only a path that considers the three areas together in an integrated fashion is sustainable.

Islands have embraced sustainable development from an early stage. The 1994 Barbados Programme of Action for the Sustainable Development of Small Island Developing States (UNGA 1994), followed by the Mauritius Declaration and Samoa Pathway, in 2005 and 2014, respectively (United Nations 2005; UNGA 2014), put sustainable development front and centre on the political agendas of Small Island Developing States (SIDS) (Baldacchino and Kelman 2014, p. 4; Douglas 2006).

The pertinent question for the purposes of this chapter is the extent to which biodiversity protection can enhance both resilience and sustainability. The short answer in each case is yes.

Resilience in the context of an environmental crisis, such as climate change or water scarcity, requires nature-based solutions (Leal Filho et al. 2020). Nature-based solutions are important for climate change adaptation and integrated resources water management. For nature-based solutions to work effectively, however, biodiversity needs to be protected and loss of biodiversity halted. Some shocks, especially to islands, are the result of attacks to biodiversity through, for example, the introduction of alien and invasive species. Clearly, for an island to bounce back (that is, to be resilient), an understanding of its biodiversity, and any threats thereto, is paramount and a different approach to biodiversity policy is needed. The policy attention given to biodiversity often needs to be integrated into wider island agendas to be effective.

Again, this is where resilience and sustainability should work together. Resilience should be part of a sustainability agenda and no agenda can be considered truly sustainable if it does not promote greater resilience.

As with resilience, sustainable development also requires paying attention to biodiversity. The very nature of sustainable development incorporates an environmental dimension and a healthy environment can be attained only by having a rich and diverse biodiversity. The SDGs set specific goals that align very closely with land and marine biodiversity. For example, SDG Goal 14.2 specifically requires the "[protection of] marine and coastal ecosystems to avoid significant adverse impact, including by strengthening their resilience". Moreover, SDG 15.4, 15.5, 15.9 and 15a all refer to the need for biodiversity protection. However, land and marine biodiversity are also essential to all the other SDGs, which are likewise crucial to meeting the biodiversity goals (Sindico 2016). An integrated approach to resilience and sustainability is required and biodiversity protection must be part and parcel of such an approach.

When such an approach is translated into policy terms, its effectiveness will also depend on the actors who shape and drive it. As explained earlier in the chapter, the CBD has opened the way to paying increased policy attention to island biodiversity and given SNIJs an opportunity to contribute meaningfully to the shaping and development of sub-national biodiversity policies and agendas. The example of the Åland Islands offers a concrete example of an SNIJ taking leadership in the development of its own sustainability agenda and recognizing the need to put biodiversity protection on that agenda.

8.5 Conclusion

Resilience and sustainability "could mean virtually anything to anyone" (Kelman and Randall 2018, p. 353). Rather than obsessing over the precise contours of a definition of resilience and sustainability that may prove more useful for dictionary purposes than for anything else, it is important to take a step back when discussing the concepts and ask whether they are understood by specific island communities. That is, do the building blocks of resilience and sustainability align with the key characteristics of island life? According to Kelman and Randall (2018, p. 355):

> The word [resilience] does not exist in their [many island cultures'] languages and is difficult to translate culturally. 'Social-ecological systems', 'coupled human and natural systems', dealing with adversity, and managing change are what many indigenous islanders have always known within their knowledge and how they have always lived: it is island life.

Indeed, some cultures and languages have no word for resilience or sustainability. This demonstrates a bias towards English as well as a western approach to society and development when pushing for global resilience and sustainability agendas. However, the example of the Åland Islands shows that key aspects of resilience and sustainability can indeed be present in local words and concepts that hew closely to the way in which island communities operate and function. In fact, the word *Bärkraft*, which is the name of the Åland sustainability network and movement, can be loosely translated as resilience (Granholm 2020).

Against this backdrop, the present chapter has explored the extent to which the CBD embraces island biodiversity as a specific area of normative concern and whether SNIJs can rely on the CBD in order to become more resilient and sustainable. On one hand, the CBD showed promising attention to island biodiversity by developing an *ad hoc* programme of work in 2006 (CBD 2006). Since 2012, however, the momentum has been lost and the CBD CoP has given no concrete attention to the island biodiversity programme of work. On the other hand, the Plan of Action on Sub-national Governments, Cities and Other Local Authorities for Biodiversity (2011–2020) (CBD 2010) has provided a much needed policy space to promote biodiversity that SNIJs can rely on to put biodiversity policies on their island agendas. The four key aspects of the Plan of Action on Sub-national Governments, Cities and Other Local Authorities for Biodiversity are present in the Åland Islands context. The Åland Islands, which have the necessary constitutional powers to take a proactive role, have employed a bottom-up approach with effective public participation to develop their sustainability agenda, which includes recognition of the need for biodiversity protection. Indeed, they can be framed as a hub of innovation whose good practices are worthy of sharing. Finally, they do not consider biodiversity protection as a stand-alone policy area, but rather as part of an integrated approach to sustainable development.

A further question raised in the chapter is the extent to which biodiversity protection is needed to promote island resilience and sustainability. In light of the scholarly debate on resilience and sustainability in island studies, the conclusion is that biodiversity policies can play a role in the wider understanding of resilience and sustainability.

I fully agree with Kelman and Randall (2018, p. 364) when they propose an agenda in the context of island studies that would go forward by addressing resilience and sustainability:

> ... [O]ne agenda item is to understand better how different island cultures and island contexts define and apply 'resilience' and 'sustainability'. In which islands' language do the words and concepts exist, how they are applied and not applied? How might islanders wish to see them being applied or avoided? How might the consequences and effectiveness of island resilience and sustainability initiatives be measured, monitored and evaluated?

The key question is who drives, develops and implements an agenda for island resilience and sustainability. Allowing SNIJs to rely on the CBD to foster biodiversity policies does not necessarily take the agenda away from the heart of the island community. In fact, SNIJs should not be oblivious of global sustainability efforts, such as the ones addressed in the SDGs. The Åland Islands epitomise the ways in which islands can take local approaches to sustainability that also consider the global context. Another very good example comes from Hawai'i (Hawai'i Green Growth 2020). Rather than discarding the SDGs, SNIJs have an opportunity to build on the synergies that may exist between the UN's intergovernmental-based understanding of sustainability and the local approach to sustainable development.

To conclude, the Åland Islands provide an example of how international legal processes and agreed objectives, such as the SDGs and to a lesser extent the CBD, can be incorporated into and drive forward resilience and sustainability agendas that include biodiversity protection and take bottom-up approaches. A bottom-up approach gives such agendas a better chance to be understood by the community and gain its necessary buy-in. Ultimately, success will also depend on other factors, such as the need to develop sounder island-specific data and indicators. However, the path to success starts as close as possible to the heart of the island community, and builds on its own understanding of resilience and sustainability.

References

Baldacchino, G., & Kelman, I. (2014). Critiquing the pursuit of island sustainability: Blue and Green, with hardly a colour in between. *Shima: The International Journal of Research into Island Cultures, 8*(2), 1–21.

Bärkraft. (2016). Development and Sustainability Agenda for Åland. Retrieved November 2, 2020, from https://sustainabledevelopment.un.org/content/documents/1523development_and_sustainability_agenda_for_aland.pdf

Bärkraft. (2017). Development and Sustainability Agenda for Åland: Status Report 1. Retrieved November 2, 2020, from https://www.barkraft.ax/sites/default/files/attachments/page/statusrapportbarkraft-en-k2.pdf

Bärkraft. (2019). *The European Sustainability Award 2019 Goes to the Development and Sustainability Agenda for Åland – The Sole Nordic Winner* [Press release]. 8 April. Retrieved November 16, 2020, from https://www.barkraft.ax/european-sustainability-award-2019-goes-development-and-sustainability-agenda-aland-sole-nordic

Begon, M., Howarth, R.W. & Townsend, C.R. (2003). *Essentials of ecology*. Oxford: Blackwell.

CBD. (1992). 1760 U.N.T.S. 69, opened for signature 5 June 1992, entered into force 29 December 1993.

CBD. (2006). CoP Decision VIII/1, *Island Biodiversity* (15 June 2006) UN Doc. UNEP/CBD/COP/DEC/VIII/I.

CBD. (2008). CoP Decision IX/21, *Island Biodiversity* (9 October 2008) UN Doc. UNEP/CBD/COP/DEC/IX/21.

CBD. (2010). CoP Decision X/22, *Plan of Action on Subnational Governments, Cities and Other Local Authorities for Biodiversity* (29 October 2010) UN Doc. UNEP/CBD/COP/DEC/X/22.

CBD. (2012). CoP Decision XI/15, *Review of the Programme of Work on Island Biodiversity* (5 December 2012) UN Doc. UNEP/CBD/COP/DEC/XI/15.

CBD. (2018a). CoP Decision XIV/34, *Comprehensive and Participatory Process for the Preparation of the Post-2020 Global Biodiversity Framework* (30 November 2018) UN Doc. CBD/COP/DEC/14/34.

CBD. (2018b). *Long-Term Strategic Directions to the 2050 Vision For Biodiversity, Approaches to Living in Harmony With Nature and Preparation for the Post-2020 Global Biodiversity Framework*, Note by the Executive Secretary (5 October 2018) UN Doc. CBD/COP/14/9.

CBD Secretariat. (2005). *Handbook of the convention on biological diversity* (3rd ed.). Montreal, Canada.

CBD Secretariat. (2020). *Global Biodiversity Outlook 5*. Montreal, Canada.

Depraetere, C., & Dahl, A. (2018). Locations and classifications. In G. Baldacchino (Ed.), *The Routledge international handbook of island studies* (pp. 21–51). Abingdon, UK: Routledge.
Douglas, C. (2006). Small island states and territories: Sustainable development issues and strategies – Challenges for changing islands in a changing world. *Sustainable Development, 14*, 75–80.
Granholm, P. (2020). *Åland Islands,* COVID-19 Island Insights Series, No. 5, University of Strathclyde Centre for Environmental Law and Governance, University of Prince Edward Island Institute of Island Studies and Island Innovation.
Hale, T. (2018). *The role of sub-state and non-state actors i6n international climate processes.* London: Chatham House.
Hawai'i Green Growth. (2020). Aloha+ Challenge. Retrieved November 16, 2020, from https://aloha-challenge.hawaiigreengrowth.org/
Johannesburg Declaration on Sustainable Development 2002, UN Doc. A/CONF.199/20.
Kelman, I., & Randall, J. (2018). Resilience and sustainability. In G. Baldacchino (Ed.), *The Routledge international handbook of island studies* (pp. 353–367). Abingdon, UK: Routledge.
Leal Filho, W., Barbir, J., Nagy, G. J., Sima, M., Kalbus, A., Paletta, A., et al. (2020). Reviewing the role of ecosystems services in the sustainability of the urban environment: A multi-country analysis. *Journal of Cleaner Production, 262*, 1–14.
Lindstrom, B. (2000). Culture and economic development in Åland. In G. Baldacchino & D. Milne (Eds.), *Lessons from the political economy of small islands: the resourcefulness of jurisdiction* (pp. 107–120). Basingstoke: Palgrave MacMillan.
Nature Scot. (2020). *Machair* [online]. Retrieved October 31, 2020, from https://www.nature.scot/landscapes-and-habitats/habitat-types/coast-and-seas/coastal-habitats/machair
Nolan, S. F. (2016). Chapter II.10 - Environmental decision making of local governments. In M. Faure (Ed.), *Elgar encyclopedia of environmental law* (pp. 136–148). Edward Elgar.
Paris Agreement 2015, UN Doc. FCCC/CP/2015/L.9/Rev1, Annex, adopted 12 December 2015, entered into force 4 November 2016.
Schroder, M. (2020). Governments fail to meet the Aichi Biodiversity Targets – Do they have a pathway towards "transformative change"?. *SCELG Policy Brief 16/2020.*
Sindico, F. (2016). *Transboundary Water Cooperation and the Sustainable Development Goals,* UNESCO–IHP Advocacy Paper.
Sindico, F. (2017). Islands, land of innovative solutions for all territories. *SCELG Dialogue 4/2017.*
Sindico, F., & Crook, N. (2021). The Islands (Scotland) Act: Island proofing through legislation. In K. Burnett, R. Burnett, & M. Danson (Eds.), *Scotland and Islandness: Explorations in community, economy and culture*. Peter Lang. Publication forthcoming.
UNFCCC. (1992). 1771 UNTS 107, opened for signature 4 June 1992, entered into force 21 March 1994.
UNFCCC. (2011). COP Decision 1/CP.16, *The Cancun Agreements: Outcome of the Work of the Ad Hoc Working Group on Long-Term Cooperative Action under the Convention* (15 March 2011) UN Doc FCCC/CP/2010/7/Add.1.
UNFCCC. (2015). COP Decision 1/CP/21, *Adoption of the Paris Agreement* (29 January 2016) UN Doc. FCCC/CP/2015/10/Add.1.
UNGA. (1994). *Report of the Global Conference on the Sustainable Development of Small Island Developing States*, UN Doc. A/CONF.167/9.
UNGA. (2014). *SIDS Accelerated Modalities of Action (SAMOA) Pathway* (15 December 2014) UN Doc. A/RES/69/15.
UNGA. (2015). *Transforming Our World: The 2030 Agenda for Sustainable Development* (21 October 2015) Un Doc. A/Res/70/1.
United Nations. (2005). *Mauritius Declaration*, 14 January, Un Doc. A/CONF.207/L.6.
Warrington, E., & Milne, D. (2018). Governance. In G. Baldacchino (Ed.), *The Routledge international handbook of island studies* (pp. 173–201). Abingdon, UK: Routledge.
World Commission on Environment and Development. (1987). *Our common future*. OUP.

Chapter 9
The Arctic Ocean: Ecosystem Approach in a Context of Extreme Vulnerability

Mar Campins Eritja

Contents

9.1 Introduction 158
9.2 The Arctic Council and Ecosystem Approach 162
9.3 The Effectiveness of the Arctic Council's Deployment of an Ecosystem Approach 164
9.4 How Other International Regulations May Help to Enhance an Ecosystem Approach in the Arctic Ocean 167
9.5 Conclusion 170
References 172

Abstract Climate change is exerting a profound impact on the biodiversity and the ecosystems of the Arctic region. Economic activity in the area is increasing and new threats have emerged that put the Arctic's ecosystems at risk. The application of an ecosystem approach takes on special relevance in areas like the Arctic, in which the relationships between habitats, biodiversity and human activity are particularly complex. An ecosystem approach, in its many forms, aims to offer a holistic and integrated view of the interactions between natural and human systems. However, from an international legal perspective the concept suffers from significant shortcomings. In this context, this chapter aims to outline how an ecosystem approach is embedded in the marine ecosystem and biodiversity management strategies led by the Arctic Council, and to assess the extent to which other applicable international legal tools might strengthen its implementation in a region in which the jurisdictional regime is so compartmentalized. Section 9.1 presents a brief description of the Arctic ecosystems and highlights the characteristics that render them particularly vulnerable. Section 9.2 focuses on reviewing the work carried out by the Arctic Council mainly through the Protection of the Arctic Marine Environment (PAME) and the Conservation of Arctic Flora and Fauna (CAFF) working groups, with the aim of ascertaining to what extent further development of the Arctic Council's instruments is necessary. Section 9.3 presents an overview of the application of an ecosystem

M. Campins Eritja (✉)
Universitat de Barcelona, Faculty of Law, Barcelona, Spain
e-mail: mcampins@ub.edu

M. Campins Eritja, T. Fajardo del Castillo (eds.), *Biological Diversity and International Law*, https://doi.org/10.1007/978-3-030-72961-5_9

approach by the United Nations Convention on the Law of the Sea (UNCLOS) and by few other conventions of marine environmental nature implemented by the International Maritime Organization (IMO) and the OSPAR Commission, both of them sharing the Permanent Observer status in the Arctic Council, and discusses how they might help to promote the application of an ecosystem approach in this polar area. Section 9.4 concludes the chapter with some final considerations.

9.1 Introduction

Arctic ecosystems are essential components of the world's biodiversity. The Arctic Ocean is a unique and highly vulnerable environment where the sea ice has a huge effect on the biological cycle of organic carbon and other elements. The Arctic snow cover is a key factor, since it creates a dynamic habitat that serves as a medium of transmission between microorganisms, plants, animals, nutrients, the soil, the water and the atmosphere (Larose et al. 2013). In spite of its extreme climatic conditions, the Arctic region possesses a rich and varied fauna: its marine and terrestrial ecosystems are home to more than 21,000 species of mammals, birds, fish, invertebrates, plants and fungi highly adapted to the cold, including iconic species such as the polar bear and the narwhal. The physical and chemical conditions of the sea ice provide varied habitats for microbial communities and meiofauna. Other lesser-known microorganisms maintain the integrity of the services provided by Arctic ecosystems (Arctic Council 2013a).

One of the most pressing problem in the region today is climate change (Arctic Council 2004a). The loss of the ice sheet is accelerating, and since the early 1980s has progressively increased the temperature of permafrost, contributing more than 40% of the rise in the sea level (Arctic Council 2011). The 2018 Intergovernmental Panel on Climate Change (IPCC) special report recorded the largest temperature increase in the region in the last 30 years and posits that the Arctic Ocean will be ice-free at least once every decade with an increase in global warming of 2 °C, compared with once a century if the increase is limited to 1.5 °C (IPCC 2018, pp. 178, 204–207).

The higher temperatures are causing significant changes in the functioning of Arctic ecosystems (IPCC 2019). The 2016 Arctic Resilience Report (Carson 2016) identified some of those changes that may push systems past a tipping point, beyond which it will be almost impossible to return to the current state of the environment. In addition to the loss of sea ice, they include alterations in thermohaline water circulation and acidification of the ocean, a change from polar to temperate primary production patterns, hypoxia of sea water, coastal marine eutrophication, and changes in the trophic level of ecosystems. Moreover, since the mid-twentieth century, a variety of persistent pollutants originating from outside the region have bioaccumulated in the Arctic reaching levels that threaten human health and ecosystems. Marine litter is another important problem, as the Arctic Ocean is an easily accessible point due to wind patterns and currents.

At regional level, the Arctic Council (Arctic Council 1996) is the cornerstone of the Arctic Ocean protection system (Koivurova et al. 2009; Rotten 2020; Barry et al. 2020a), even though it has no constituent treaty or binding decision-making powers. The Arctic Council is responsible for environmental protection and sustainable development in the Arctic Ocean. Besides its Ministerial Meetings, it achieves this objective through the Senior Arctic Officials—SAO—, the working groups, the task forces and the expert groups. Particularly important is the work done by its six specific working groups: the Protection of the Arctic Marine Environment (PAME), the Conservation of Arctic Flora and Fauna (CAFF), the Arctic Monitoring and Assessment Programme (AMAP), the Emergency Prevention, Preparedness and Response Working Group (EPPR), the Sustainable Development Working Group (SDWG), and the Arctic Contaminants Action Program (ACAP). Since 2013, it has also had a Secretariat.

The Arctic Council comprises the five Arctic coastal States, e.g., Canada, Greenland/Denmark, Norway, the Russian Federation and the United States (known as the Arctic Five) and the three other circumpolar States, Finland, Iceland and Sweden. As climate change facilitates access to Arctic resources, the region is attracting the interest of many non-Arctic actors (Young 2016). Therefore, alongside these powerful insiders, certain other players from outside the region enjoy Permanent Observer status and may clearly influence its future development (e.g., China, Japan, India, Korea, Singapore, Switzerland and several European Union member States). However, the evolution of this region is entirely dependent on the national strategies of the Arctic Five (Barry et al. 2020a).

Among the many challenges facing the Arctic Ocean, some of the most significant are the increase in maritime traffic, the possibility of drilling oil, gas and mineral reserves, and access to new fisheries. All these developments have major political and economic implications. On the one hand, maritime traffic has grown by 25% between 2013 and 2019. This traffic mostly comprises fishing vessels, but there is also a growing presence of other types of ships and vessels (Arctic Council 2020). On the other hand, there is a considerable geostrategic interest in Arctic mineral resources, since the region is home to around 13% of the world's reserves of oil and 30% of its reserves of gas (Bird et al. 2008; Gautier et al. 2009). Although activities related to offshore hydrocarbon drilling in the Arctic are still of secondary importance in the global context, and in fact have been decreasing in recent years, demand for these resources may well rise unless a genuine energy transition takes place (Osofsky et al. 2016). Finally, the risk of fishery overexploitation and the presence of illegal, unreported and unregulated fishing takes on a new dimension in the Arctic Ocean (Christiansen et al. 2014; ICES 2018). According to the agreement reached by the Arctic Five in the 2008 Ilulissat Declaration (Arctic Ocean Conference 2008), these challenges are to be managed in accordance with existing international law, specifically through the United Nations Convention on the Law of the Sea (UNCLOS) (UNCLOS 1982).

In this scenario, the application of an ecosystem approach is especially relevant in an area like the Arctic, in which the relationships between habitats, biodiversity and human activity are particularly complex.[1] In general, an ecosystem approach, which has traditionally been associated with the marine environment, aims to manage the use of biodiversity and ecosystems by considering, on the one hand, the potential damage that human activities can cause to them and, on the other, the importance and complexity of ecological relationships. Thus, it assesses the interactions between natural and human systems in a holistic and integral way (Trouwborst 2009; Langlet and Rayfuse 2019). The problem (as is usually the case) lies in the details and, in particular, in identifying the requirements and the core features of this approach, since most of the international instruments that adopt it do not describe them. The Conference of the Parties (CoP) of the Convention on Biological Diversity (CBD) (CBD 1992) has defined an ecosystem approach as "a strategy for the integrated management of land, water and living resources that promotes conservation and sustainable use in an equitable way" (CBD 2000, 2004b). As noted by De Lucia, the CBD offers a nuanced definition, in which the ecosystem approach is a holistic concept that "is first understood in relation to ecosystems as objective entities" and, simultaneously, is "aimed at managing natural resources for the long-term benefit of humanity" (De Lucia 2015, p. 111). Besides, the concept is closely linked to the twelve Malawi Principles (CBD 1998, 2004c), which promote an approach to decision-making that considers how ecosystems are affected in space and time and incorporate the need to recognize the participation of stakeholders.

The rationale of an ecosystem approach underpins all the CBD's programmes, but its scope remains unclear. In its various forms and supported by multiple modelling tools,[2] it shows great potential for contributing to the sustainable management of biodiversity—especially when operating in a given spatial framework. It provides reliable scientific advice on the dynamics and interactions between ecosystems and human activity and, when related to the specific goals of biodiversity and ecosystem management policies, it facilitates the adoption of recommendations or decisions by public authorities, so as to optimize the use of the resource ecologically, economically or socially.

However, from an international legal perspective, it suffers from significant shortcomings and is weak from the outset due to its poorly established legal basis.

---

[1]While there is a considerable amount of academic literature on this approach from an international law perspective (see, for all, Trouwborst 2009), little attention has been paid to its implementation, and especially to its practical implications for the development of sustainability policies in the Arctic (McDonough et al. 2017; Malinauskaite et al. 2019). The use of multiple terms to refer to ecosystem approaches is common, and leads to a certain "Babylonian confusion" (Trouwborst 2009, p. 28, Kirk 2015). However, while each of them has its own key implications, for the purposes of this chapter the concept is used in its generic sense, regardless of how the different legal frameworks discussed refer to it.

[2]For example, the Arctic Biodiversity Data Service (ABDS) facilitates long-term data for modelling terrestrial, freshwater and marine Arctic ecosystem-based management: http://geo.abds.is/geonetwork/srv/cat/catalog.search#/home.

Within the existing international law,[3] the concept is difficult to grasp—especially the question of whether it is merely a policy approach or is a narrative within the broader one of sustainable development (De Lucia 2015). Briefly, deriving concrete legal effects from this approach remains a challenge since it needs to be linked to principles which are still being consolidated themselves (Koester 2018). Its evolving nature also adds to the confusion about its meaning (Enright and Boteler 2020) and it remains elusive and unstable (Koester 2018; De Lucia 2015; Trouwborst 2009). For this reason, the complementary Malawi principles also suffer from a significant degree of vagueness and their application entails considerable difficulty (Waylen et al. 2013). As such, this set of principles only provide a "basic frame of reference for State Parties" (De Lucia 2015, p. 111). However, as is clear from documents produced by the CBD's CoP, strict adherence is probably not as important as the holistic philosophy they imply (CBD 2006), regardless the little attention it has received or its weak influence on other initiatives (Waylen et al. 2014).

In addition to its restricted legal impact, the main challenge facing the ecosystem approach lies in its practical application, which may occur in different ways or be carried out to different degrees. Ultimately, it does not really guarantee effective management of ecosystems *per se*, but aims to manage human activities taking ecosystems into account. Insofar as it is only a methodology for establishing management or policy priorities in certain areas or ecosystems, only when it is translated into an operational guidance or a legally binding instrument does it have any impact on the actions of States.

Ultimately, an ecosystem approach needs to be implemented through specific tools. It is significant that while the CBD's bodies still find it difficult to prescribe the specific elements of this approach, they have made much faster progress with respect to other interacting tools that clearly serve this purpose. This is the case, for example, of protected areas. The CBD plays a central role in the development of this traditional *in situ* instrument (CBD 2003, 2004a) and one of its main contributions is the recognition of the Ecologically or Biologically Significant Areas (EBSAs).[4] So far, the CoP of the CBD has identified 11 EBSAs in the Arctic Ocean (CBD 2014, 2015; Jakobsen 2016), two of them located in the Central Arctic Ocean and nine others under the jurisdiction of the Russian Federation.

The question that is addressed in this chapter is how an ecosystem approach is embedded into the marine ecosystem and biodiversity planning and management strategies led by the Arctic Council and to what extent other applicable international legal tools can strengthen its implementation, in a region in which the jurisdictional regime is so compartmentalized. Therefore, Sect. 9.2 focuses on reviewing the work carried out by the Arctic Council mainly through the PAME and CAFF working

[3]The first reference to the ecosystem approach is found in the 1980 Convention on the Conservation of Antarctic Marine Living Resources (1982, 19 ILM 841). Article I.3 includes the concept of "Antarctic marine ecosystem" and Article II.3.c requires that all conservation activities be carried out with respect to the "marine ecosystem as a whole", in order to allow the sustained conservation of Antarctic marine living resources.

[4]See Chap. 11.

groups, in an attempt to determine the extent to which further development of the Arctic Council's instruments is necessary. Section 9.3 presents an overview of the application of this ecosystem approach by the UNCLOS and by other conventions of a marine environmental nature implemented by the International Maritime Organization and the OSPAR Commission (i.e., a global organization and a regional one) both of which have Permanent Observer status in the Arctic Council, and discusses how they might help to promote the application of an ecosystem approach in this polar area. Section 9.4 concludes the chapter with some final considerations.

9.2 The Arctic Council and Ecosystem Approach

The CBD is a key instrument for the promotion of biodiversity and ecosystems conservation, especially within national jurisdictions. Its provisions and the Decisions of its CoP have had an effect that has clearly spilled over into other related environmental regimes. With the CBD as the point of departure, this is the case for the Arctic region. However, the Arctic Ocean's biodiversity received only marginal attention from the CBD bodies until 2010, when its CoP recommended its "consideration as a new and emerging" issue (CBD 2010). A first partnership agreement on cooperation between the CBD Secretariat and the Arctic Council was signed in 2009, followed by the invitation by the CBD to the Arctic Council for CAFF to undertake periodic assessments that would enable the CBD's subsidiary bodies to evaluate the state of Arctic biodiversity. In the following year, the Parties were urged "to promote the implementation of the Strategic Plan for Biodiversity 2011-2020 and relevant programmes of work (. . .) in relation to the Arctic environment" (CBD 2011). This action resulted in the Arctic Biodiversity Assessment produced by CAFF in 2013, the only one published so far. This assessment allowed the adoption of a set of 17 policy recommendations, among which the third calls for progress in the efforts to implement an ecosystem-based approach as a framework for cooperation, planning, and development (Arctic Council 2013a).

The marine environment has traditionally received more attention within the Arctic Council, a circumstance that explains the prominent role of PAME in the building of this ecosystem approach. PAME was central to the development of the first Arctic Marine Strategic Plan in 2004, renewed in 2015 for the period 2015–2025 (Arctic Council 2015a, b).

In that document, PAME endorsed the CBD's ecosystem approach as one that "requires that development activities be coordinated in a way that minimizes their impact on the environment and integrates thinking across environmental, socio-economic, political and sectoral realms." (Arctic Council 2004a). Some years later, PAME defined it in the Arctic Ocean Review Project as

> the comprehensive integrated management of human activities based on best available scientific knowledge about the ecosystem and its dynamics, in order to identify and take action on influences that are critical to the health of ecosystems, thereby achieving

sustainable use of ecosystem goods and services and maintenance of ecosystem integrity (Arctic Council 2013b).

This more elaborated approach involves careful consideration of the options and trade-offs associated with different management options, and the Arctic Council has devoted considerable efforts to consolidating its development and identifying the underlying elements that should support the decision-making process (Arctic Council 2015a, 2017).

With this aim, PAME established an expert group to develop a series of guidelines for an ecosystem approach to marine management in 2007, which was opened to the other working groups in 2011. This task, which led to the drafting of the guidelines adopted by PAME in 2019 (Arctic Council 2019), focuses on identifying the ecosystems, describing the biological and physical components and processes, assessing the state of the ecosystems, valuing their goods and services, and managing human activities therein. All these elements were addressed in successive PAME workshops and conferences.

Moreover, the 2004 Arctic Marine Strategic Plan strengthened the impetus for implementing the ecosystem approach, while recognizing its importance for the conservation of marine biodiversity and Arctic ecosystems. To this end, it proposed a series of actions, one of the most important being the recognition of Arctic Large Marine Ecosystems (LMEs). Accordingly, PAME identified a list of 18 major marine areas in Arctic and sub-Arctic waters in 2013 (Arctic Council 2013c; Jakobsen 2016). One of these LMEs corresponds in practice to the Central Arctic Ocean and is the only area partially outside national jurisdiction. Thirteen other LMEs are located in areas under Icelandic, Norwegian, American, Russian and Canadian jurisdiction, and four are under the jurisdiction of France and Denmark. In parallel, the PAME identified a total of 97 additional areas as areas of major ecological and cultural importance (Arctic Council 2013d).

Amongst PAME's objectives (Arctic Council 2015b), the establishment of a Framework for a Circumpolar Network of Marine Protected Areas (FCNMPA) (Arctic Council 2015c) is particularly noteworthy for the implementation of an ecosystem approach. The FCNMPA sets out four main objectives that have driven its development since then: strengthening ecological resilience; sustaining cultural, social and economic values and ecosystem services; improving public awareness and support; and promoting coordination and collaboration among Arctic States in the planning and management of protected areas (Arctic Council 2015a).

On the other side, CAFF is the main working group responsible for transmitting information on the state of the Arctic's biodiversity to the CBD's bodies as well as to the Intergovernmental Science Policy Platform on Biodiversity and Ecosystem Services (IPBES) (Barry et al. 2020b).

At the request of the Arctic Council's ministers, it developed a series of draft guiding principles for biodiversity management (Arctic Council 2015a) based on an ecosystem approach. As a starting point, CAFF considers "all the functions, processes, interactions, and life stages that make up an ecosystem, not just individual species or resources, and planning for and adapting to both expected changes and unexpected events", the guidelines being also "holistic and consistent with the perspectives of Arctic Indigenous Peoples" (Arctic Council 2016). On that assumption, CAFF has devised various programmes. On the one hand, the Circumpolar Biodiversity Monitoring Programme (CBMP), an international network to report on changes in Arctic biodiversity, has promoted the establishment of dedicated working groups and four ecosystem-based monitoring plans (marine, freshwater, terrestrial and coastal),[5] each of them identifying the focal ecosystem components in which changes probably indicate changes in the global environment. On the other hand, the Circumpolar Protected Areas Network (CPAN) serves to coordinate States' policies regarding protected areas in their Arctic regions and highlights their value not only in scientific but also in educational, cultural, recreational and spiritual terms.

9.3 The Effectiveness of the Arctic Council's Deployment of an Ecosystem Approach

The ecosystem approach promoted within the CBD has filtered into the activities of the Arctic Council, especially since 2004. In principle, the existence of close relations between the ecosystems in the Arctic region, and the possibility of defining more precise spatial and ecological boundaries between them, may facilitate the effective incorporation of an ecosystem approach. However, it seems that there is still much room for improvement. Although there has been some significant progress, the ecosystem approach can hardly provide relevant elements for the decision-making process and it still has a marginal impact in that context. In the end, this debate highlights the difficulty of balancing highly complex ecological requirements with those of socio-economic systems and with the different perspectives and values of the stakeholders concerned.

There are several possible reasons that go some way to explaining the difficulties in its deployment. On the one hand, the development and implementation of an ecosystem approach in the Arctic Ocean still is highly conditioned by scientific uncertainty and requires the consideration of a vast region, spanning the temperate waters of the northern part of the North Atlantic Ocean and the North Pacific Ocean and the areas covered by permanent ice in the Central Arctic Ocean. These areas present important differences in terms of their natural conditions and a great diversity of ecosystems which vary from boreal to polar, and a little is still known about their reaction to ongoing environmental changes. On the other hand, the application of an ecosystem approach should also encompass the management of human activities, which in the Arctic region also differ in terms of their development and scope, ranging from subsistence level activities in the case of indigenous communities to large-scale commercial exploitation (Jentoft and Bjorkan 2007; Hoel 2010).

[5](https://www.caff.is/cbmp-monitoring-plans).

Besides, the management of the Arctic Ocean based on an ecosystem approach demands a holistic vision that international law so far seems ill-equipped to provide. Several binding international instruments that apply to this area reflect an ecosystem approach and have developed slight adaptations. Nevertheless, we find in the Arctic Ocean a set of international instruments that coexist but from which it is difficult to say that they constitute a fully integrated regime.[6]

In addition to the bilateral agreements between Arctic coastal States, there is a considerable number of international conventions covering this marine region. Some are comprehensive multilateral environmental agreements, also applicable to the Arctic (e.g., UNCLOS); others reflect explicit efforts to address specific environmental conditions arising in the Arctic, such as the 2013 Agreement on Cooperation in Marine Oil Pollution, Preparedness and Response in the Arctic (Arctic Council 2013e); and still others are regional agreements whose scope extends to the North Pole, such as the Convention for the Protection of the Marine Environmental of the North-East Atlantic (OSPAR Convention) (OSPAR 1992). In that sense, the international legal tools governing the Arctic region, even if complementary, are still fragmented (Young 2016; Barry et al. 2020a) and often built on the basis of legally non-binding provisions drawn up by the bodies of environmentally-related conventions.

Likewise, the national legal systems of the Arctic coastal States offer different ways of approaching the management of ecosystems and may have a different understanding of the ecosystem approach. From a jurisdictional point of view, the immediate coastal areas of the Arctic Ocean are included in the territorial seas of the five Arctic coastal States. Although in the Arctic coastal States sufficient legislation is in place to protect the marine environment and conserve its biodiversity and ecosystems, national regulations are heterogeneous and important gaps still remain in its application. An example is found again in the diversity of measures related to protected areas included in the national Arctic networks, all of which present constraints and are difficult to manage as a whole (Jentoft and Bjorkan 2007; Lalonde 2013; Jakobsen 2016). In this regard, considerations about the implementation of the ecosystem approach are usually highly dependent on local, sub-national or national circumstances, as they must eventually be incorporated and translated into specific policies and planning instruments. It is therefore essential to design a legal framework of wide coverage with a prioritization of objectives in order to develop and make the ecosystem approach operative.

Other problems have also arisen due to the inherent limitation of Arctic Council's working groups. First, due to the absence of binding character and the consensual nature of the measures decided by the working groups, their activity has mainly been limited to certain forms of soft power, like the creation of policy recommendations and advice for the management, creation and dissemination of knowledge on biodiversity and Arctic ecosystems, and seeking formulas to facilitate engagement between different stakeholders (Barry et al. 2020b). Second, even though PAME and

[6]See Chap. 3.

CAFF have played an important role in developing cooperative initiatives, the impetus depends on the leadership of the rotatory Chairmanship, since there is no overall Arctic Council strategy. Third, the working groups also lack any mechanisms to supervise and follow up their outcomes, wherefore the implementation of their guidelines or recommendations remains, too, in the hands of the Arctic States (Hossain and Morris 2017). Fourth, the working groups have also suffered from the Arctic Council's limited resources and the lack of a programming budget (Barry et al. 2020a). As the funding of their activities is voluntary and must be provided by States or other entities, long-term projects are difficult to design and manage. Ultimately, the functioning of the working groups tends to be excessively influenced by the interests of the Arctic Council States (Rotten 2020; Barry et al. 2020a).

As for PAME, the work of the joint expert group for an ecosystem approach to marine management which began in 2007 only started to show results in 2019, having had to face "practical issues of an often expensive, time-consuming and labour-intensive field work" (French 2017) which slowed the crystallization of tangible outcomes. In contrast, PAME has played a more important role through the Arctic Marine Strategic Plan and the establishment of a FCNMPA. Though, despite the fact that this was a priority action and has undoubtedly been its most ambitious project, PAME has taken few concrete steps to implement the Circumpolar Network and progress to date has been limited (Koivurova 2009; Hossain and Morris 2017; Polakowski 2017). In fact, the FCNMPA, built on the networks of national protected areas, is at present no more than a "piece-by-piece coordination of nation-led efforts to create MPAs [States' protected areas] within sovereign borders" (Hossain and Morris 2017, p. 107). Nor does it contain any substantive commitment for the members of the Arctic Council with regard to the extent, scope or establishment of national protected areas, since, as it warns from its very first pages, it only provides guidelines that members can use in accordance with their own objectives and interests (Arctic Council 2015c). However, PAME also carries out other tasks related to the management of Arctic LMEs and to the development of best practices in ecosystem ocean management, which can provide a basis for improving the potential of the Circumpolar Network (Lalonde 2013; Allen 2018).

As far as CAFF is concerned, almost eight years after the adoption of the Arctic Biodiversity Assessment, most of the recommendations made at that time have yet to be fully implemented. Likewise, although in the early years the CPAN was also presented as CAFF's highest priority, its activity has waned significantly since 2004, mainly due to the low level of involvement of the Arctic coastal States. However, the CPAN still makes a clear contribution to the adoption of sustainable development measures by supporting the attempts of the Arctic Council's States to fulfil the obligations imposed on them by other international instruments. For its part, the CBMP seems to have made significant progress, probably because it is more in line with the operational spirit of the Arctic Council and does not require an intense political commitment (Arctic Council 2005; Koivurova 2009; Lalonde 2013). Besides, the overlap with the tasks already carried out by the Arctic Council States under the CBD calls into question the need for some of the specific actions inside CAFF (Koivurova 2009).

9.4 How Other International Regulations May Help to Enhance an Ecosystem Approach in the Arctic Ocean

The Arctic Council has been responsive to the contributions that promote the implementation of an ecosystem approach, which has come to occupy a particularly prominent place in the context of marine biodiversity management. It has been increasingly adopted, directly or indirectly, in other internationally binding instruments and might support its application in the Arctic Ocean. This has occurred in instruments promoted at the universal level, or by more institutionalized frameworks enjoying Permanent Observer status in the Arctic Council.

UNCLOS did not explicitly include an ecosystem approach because, at the time of its adoption, the concept was still alien to international law. However, its updated interpretation at the light of the CBD confirms that, at least implicitly, UNCLOS also embodies this approach to marine environmental protection.[7]

To begin with, its preamble recognized that "problems of ocean space are closely interrelated and need to be considered as a whole". Under Articles 192–196, States have a general obligation to "protect and preserve the marine environment". This is a general obligation of conduct which, for the common and general interest of all States, includes a positive obligation for the States to take measures to prevent or at least mitigate damage to the marine environment, and a negative obligation not to allow activities within their jurisdiction and control to degrade it (ICJ 1996, para. 29; ICJ 2010, para. 101 and 184; PCA 2016, para. 956–959). It also extends to all maritime areas, both under the jurisdiction of the coastal States and beyond (ITLOS 2015, para. 111–116; PCA 2016, para. 941–945). Furthermore, Article 194.5 allows special treatment for rare and fragile ecosystems and the habitats of "depleted, threatened or endangered" species. It also establishes the obligation of States to take all necessary measures to prevent, reduce and control marine pollution from all sources (art. 194.1), with a particular reference to those resulting from the use of technologies or the introduction of species (art. 196). Furthermore, several provisions recognize the existence of particular circumstances that justify greater legal protection in certain areas, which might provide a legal basis for the special treatment of the Arctic Ocean's ecosystems (art. 195.5).

Apart from this, although Article 211.6 of UNCLOS was not initially intended to apply to areas with the characteristics of the Arctic Ocean, the Arctic presents several criteria that qualify it for inclusion in the category of "particular and clearly defined area" within the Exclusive Economic Zone (EEZ) which requires "special mandatory measures to prevent pollution" and in which the application of an ecosystem approach is even more justified: the oceanographic characteristics, the presence of sea ice, the climate, and the ecological conditions, since threatened marine species and particularly fragile ecosystems coexist with increasingly large commercial fisheries (Sebek 1990; Van Dyke and Broder 2012; Jakobsen 2013, 2016).

[7] See Chap. 2.

Besides, a broad interpretation of Article 234 (the "Arctic exception") might allow its consideration as a special rule as opposed to Article 211.6. This provision, which has been invoked, for example, by Canada following its acceptance of the optional annexes to the Convention for the Prevention of Pollution from Ships (MARPOL Convention) (IMO 2020a), is not unambiguous, since the conditions it sets do not precisely define either the circumstances or the area in which the more restrictive measures may be applied. On the assumption that this provision aims to establish additional guarantees for the marine sustainable use, a broader teleological and contextual interpretation might identify the "ice-covered areas" as the particularly vulnerable area of Arctic maritime spaces (Bartenstein 2011; Jakobsen 2013; McRae and Goundrey 1982; Molenaar 2009; De Lucia 2017; Rayfuse 2014; Gavrilov et al. 2019).

Nevertheless, even if UNCLOS is considered to have adopted an ecosystem approach through the generalized obligation of marine environment protection, and indeed incorporates some of its attributes, it does so only implicitly and on the basis of a broad interpretation of its literal text. Beyond this, it is difficult to establish how it should be articulated.

Other conventions of environmental nature implemented by the International Maritime Organization (IMO) and the OSPAR Commission, both of them sharing the Permanent Observer status in the Arctic Council, might help to promote the application of an ecosystem approach in this polar area.

IMO was granted Permanent Observer status in 2019, although this collaboration had already been taking place for some time. Thus, in 2016, IMO developed jointly with the EPPR the Guide on Response to Oil Spills in Snow and Ice Conditions (IMO 2017b). The Polar Code approved by IMO in 2015 (IMO 2015a) also required intense cooperation. Besides, IMO cooperates closely with PAME and deals with a number of issues relating to maritime transport in the Arctic, in particular search and rescue, combating pollution, maritime safety and protection of the marine and atmospheric environment.

The IMO's MARPOL Convention (MARPOL 1973/1978), ratified by all Arctic coastal States, sets international standards for the management of discharges from ships of oil and other substances harmful to the marine environment. The obligations of the parties are set out in six annexes which, as of 2019, have been accepted by all eight Arctic States with the exception of the United States in respect of Annex IV (IMO 2020a).

Although MARPOL Convention does not mention an ecosystem approach, it makes use of the traditional tool closely related to it, the protected areas. Annexes I, II and V recognize the existence of Special Areas (SAs) that require greater marine environmental protection, thus imposing specific restrictions. So far, no SAs have been designated in the Arctic Ocean (IMO 2018). However, even if the Arctic Ocean is not expressly mentioned (as are, for example, the Mediterranean Sea, the Baltic, the Black Sea, the Red Sea, Gulf areas and the Antarctic Zone), it satisfies at least the first two conditions (and also the third) required for designation as a SAs: namely, oceanographic conditions, ecological conditions and characteristics of maritime traffic (Arctic Council 2013b).

Closely related to MARPOL Convention, since it implies amendments to its Annexes I, II, IV and V, the Polar Code (IMO 2015a, b), adopted as the first set of international regulations specifically addressing potential hazards to the Arctic and Antarctic environment, may therefore also affect the designation and management of these SAs. It is worth mentioning the 2010 amendment of MARPOL Convention's Annex I (IMO 2010), which, initially designed to apply only to the Antarctic area, is of particular interest for the future of the Arctic Ocean. This amendment, which prohibits the carriage of heavy oil in the Antarctic Ocean area, was fully endorsed by the Polar Code which recommends that the same ban should also apply in the Arctic Ocean. However, due to the dissimilarities in their international legal systems, and to the great differences in navigation flows and activities in the two polar areas, the issue seems to be much more contentious in the Arctic.

Despite these challenges, the task currently being carried out by the IMO Marine Environmental Protection Committee allows for a certain amount of optimism with regard to the new MARPOL Convention's guidelines for navigation, which include the prohibition for the use and carriage of heavy fuel oil in Arctic waters in July 2024 (IMO 2017a). Its Sub-Committee on Pollution Prevention and Response has already discussed the possibility of a prohibition of this kind (IMO 2020b), although the ban does not seem to go far enough since it does not apply to black carbon (IMO 2015c, 2019), and, in contrast to the Antarctic area, allows large-scale exceptions and waivers until July 2029 for ships sailing under the flags of Arctic coastal States (Comer et al. 2020). Despite the current shortcomings and loopholes, nevertheless, some progress within the IMO's bodies, albeit limited, points to a certain optimism.

Finally, even though it is a regional agreement, it is worth mentioning here the OSPAR Convention—applicable to the North Pole in the European Arctic—and the role of the OSPAR Commission, which was also granted Permanent Observer status to the Arctic Council in 2017. Since then, it has been actively involved with AMAP and PAME in the field of marine environmental protection issues.

OSPAR Convention applies to approximately one quarter of the Arctic and sub-Arctic waters (Region I). A major shortcoming is that it does not cover all uses of the oceans that may be relevant to ecosystem approach management, which is a clear limitation for an effective implementation (Smith and Jabour 2018). It does not extend, for example, to fisheries or maritime transport. To fill this gap, even if only partially, Annex V "On the protection and Conservation of the Ecosystems and Biological Diversity" provides for cooperation between the OSPAR Commission and the competent authorities under other conventions such as the North- East Atlantic Fisheries Commission (NEAFC), the IMO or the UNCLOS International Seabed Authority.

Article 2.1 of OSPAR Convention establishes the obligation for the Parties to prevent pollution and preserve marine ecosystems, and Annex V, which broadens the OSPAR Convention at this point (Jakobsen 2016), requires them to cooperate in the adoption of measures to protect and conserve marine biological diversity as well. The OSPAR Convention does not expressly mention the concept of an ecosystem approach, but its assumption is implicit in the Preamble, the article 2 and the Annex V. To confirm it, the OSPAR Commission bases its task on "an integrated ecosystem approach", defined as "the comprehensive integrated management of human activities based on the best available scientific knowledge about the ecosystem and its dynamics, in order to identify and take action on influences which are critical to the health of marine ecosystems, thereby achieving sustainable use of ecosystem goods and services and maintenance of ecosystem integrity" (OSPAR 2003). It has also clearly endorsed the ecosystem approach as an "overarching principle" to be applied to the management of all human activities that have an impact on the needs of the marine environment in its Strategy for the Protection of the Marine Environment 2010–2020 and is fully committed to its progressive implementation (OSPAR 2010b).

Turning again to one of the traditional instruments for the implementation of the ecosystem approach, the OSPAR Commission has been particularly active in the designation of the Priority Marine Habitats (PMHs) and the creation of an "ecologically coherent network of Marine Protected Areas" (OSPAR 2010a). Of the two Arctic coastal States that are also OSPAR Parties—Norway and Denmark (Greenland)—only Norway has designated five areas under its jurisdiction with this international status within the Arctic Circle, in addition to five areas outside it but inside Region I. Other Parties bordering the North-East Atlantic but whose jurisdiction also extends to OSPAR's Region I, such as Iceland and the United Kingdom, have also designated protected areas (OSPAR MPA Database). No areas have been established in the Arctic high seas: a proposal was put forward in 2015, but it failed because of the reluctance of the coastal States (Jakobsen 2016, p. 231). In addition, the OSPAR Commission has been involved in the identification of eleven potential Arctic EBSAs under the umbrella of the CBD—two of them, as mentioned above, in the Central Arctic Ocean. Even so, the OSPAR's Arctic region still has the fewest protected areas (De Santo 2018, p. 38). Hence, it can be concluded that the direct involvement of the OSPAR Commission might be a significant precedent for multilateral cooperation (Hossain and Morris 2017; Molenaar and Oude Elferink 2009).

9.5 Conclusion

There is a general consensus on the need to consider the impact of human activities as part of the dynamics of Arctic Ocean's ecological systems. An ecosystem approach, especially with regard to the marine environment, makes such a consideration possible.

At the same time, there is also a widespread recognition of the difficulties that its implementation entails, due to its complexity and the lack of instruments available. It is not enough for this approach to be discussed in the framework of sustainability policies and strategies; it must be fully integrated inside them, which can be done, at least partly, through protected areas. Nevertheless, it remains a challenging concept, not only because of the scientific complexity and still limited understanding of the functioning of Arctic Ocean's ecosystems, but also because it relies heavily on the legal and institutional systems within which biodiversity and ecosystem management are framed, both at international and at state level.

Despite its intricacy, the implementation of an ecosystem approach has been considered as a priority for addressing the CBD's objectives. This consensus is still valid today, as evidenced by its inclusion in the discussion of the *Zero Draft* of the Post-2020 Global Biodiversity Framework (CBD 2020), above all as an approach that contributes positively to fisheries management, climate change mitigation and adaptation, or the restoration of ecosystem services and functions. The embedding of an ecosystem approach in the CBD is therefore an important step forward for the protection of Arctic ecosystems; it implies the recognition of their vulnerability and facilitates a response to the impacts at institutional level.

As the key regional organization, the Arctic Council clearly has an important role to play in promoting and implementing an ecosystem approach. In this regard, the initiatives undertaken by the Arctic Council's working groups in the last decade should be seen in a very positive light, as they have resulted in better monitoring of the state of the Arctic marine ecosystems. However, decisions about what to preserve, manage, and monitor, in what way, for what purpose, and for how long, require legal structures that are stable but sufficiently flexible to be able to respond both to changes in ecosystems and to the development of human activities that affect them. It also demands a clear and careful definition of institutional responsibilities at each level of action together with the participation of the appropriate actors, which should also be accompanied by a performance assessment system, something that the Arctic Council does not offer at present.

As with the Aichi Biodiversity Targets adopted in 2010, the fulfilment of many of the objectives set for the management of these ecosystems seems to have been lost along the way. As the scientific knowledge regarding the Arctic Ocean advances, the awareness of the need for international regulation is mounting. Clearly, current international law does not provide an adequate legal framework. The increasing vulnerability of the Arctic Ocean's ecosystems calls for proposals aimed at establishing a specific legal framework (for the Central Arctic Ocean at least) that clearly sets out and deploys an ecosystem approach. Hereof, from an international law perspective, several options are open that would allow the ecosystem approach to be included in one way or another, such as the adoption of a multilateral treaty along the lines of the UNCLOS implementation agreements. Consideration has been given to the adoption of an annex referring to the Central Arctic Ocean in the future "legally binding international instrument on the conservation and sustainable use of marine biological diversity in areas beyond national jurisdiction" currently under negotiation (UNGA 2015) as well as to the recognition of such ecosystem approach (De Lucia 2019). The adoption of a specific agreement between Arctic Council States, a formula that yielded notable results in the past, is another possibility. However, although these options would allow for greater flexibility and adaptation to international commitments, their adoption seems highly unlikely at this stage.

While solutions of this kind are probably necessary, they are certainly not sufficient for the effective implementation of an ecosystem approach in the Arctic Ocean—especially when there is no single governance structure, but a patchwork of legal tools and few synergies. Therefore, it is also necessary to incorporate other instruments currently in the hands of Arctic coastal States, such as those related to land management and planning, fisheries, tourism and economic development, into the management of the ecosystems. At the same time, it should also be stressed that ecosystems do not operate only within the limits drawn by national borders nor do they fit into the compartmentalised structures of governments. The "piece by piece" management of ecosystems is not an adequate response, or at least not a satisfactory one, to the complex changes that the Arctic Ocean is undergoing today.

The data provided by the assessments of the international bodies are clear: the Arctic Ocean is no longer a region isolated from the rest of the planet and it requires a legal protection framework at all scales (including international, regional, national and local scales) to safeguard marine ecosystems that are closely interconnected and critical to the biophysical stability not only of this region but of the entire planet.

References

Allen, C. (2018). Arctic Law & Policy Year in Review: 2017. *Washington Journal of Environmental Law & Policy, 8*(1), 106–263.

Arctic Council. (1996). Declaration of the Establishment of the Arctic Council, Ottawa. Retrieved September 26, 2020, from https://oaarchive.arctic-council.org/handle/11374/85

Arctic Council. (2004a). Reykjavik Ministerial Meeting, PAME: Arctic Marine Strategic Plan. Retrieved September 26, 2020, from https://oaarchive.arctic-council.org/handle/11374/71

Arctic Council. (2004b). Declaration of the Reykjavik Ministerial Meeting. Retrieved September 26, 2020, from https://oaarchive.arctic-council.org/handle/11374/89

Arctic Council. (2005). CAFF Management Board Meeting Minutes, 1–3 February 2005. Retrieved December 18, 2020, from http://arcticportal.org/uploads/t-/9F/t-9FpbaWsOdX3RSz_UyIFw/CAFF-Board-Meeting-Helsinki-Finland-February-1-3-2005.pdf

Arctic Council. (2009). PAME: Arctic Marine Shipping Assessment. Retrieved September 26, 2020, from https://pame.is/projects/arctic-marine-shipping/amsa

Arctic Council. (2011). AMAP: Snow, Water, Ice and Permafrost in the Arctic (SWIPA). Retrieved September 26, 2020, from https://www.amap.no/documents/doc/snow-water-ice-and-permafrost-in-the-arctic-swipa-climate-change-and-the-cryosphere/743

Arctic Council. (2013a). CAFF: Arctic Biodiversity Assessment. Retrieved September 26, 2020, from https://www.arcticbiodiversity.is/index.php/the-report/chapters

Arctic Council. (2013b). PAME: The Arctic Ocean Review Project, Final Report, (Phase II 2011–2013). Retrieved August 28, 2020, from https://oaarchive.arctic-council.org/handle/11374/1219

Arctic Council. (2013c). PAME: Large Marine Ecosystems (LMEs) of the Arctic area, Revision of the Arctic LME map. Retrieved September 26, 2020, from https://oaarchive.arctic-council.org/handle/11374/61

Arctic Council. (2013d). AMAP/CAFF/SDWG: Identification of Arctic marine areas of heightened ecological and cultural significance, Arctic Marine Shipping Assessment. Retrieved September 26, 2020, from http://www.amap.no/documents/doc/identification-of-arctic-marine-areas-of-heightened

Arctic Council. (2013e). Agreement on Cooperation on Aeronautical and Maritime Search and Rescue in the Arctic. Retrieved September 26, 2020, from https://oaarchive.arctic-council.org/handle/11374/531

Arctic Council. (2015a). Declaration of the Iqaluit Ministerial Meeting. Retrieved September 26, 2020, from https://oaarchive.arctic-council.org/handle/11374/662

Arctic Council. (2015b). PAME: Arctic Marine Strategic Plan: Protecting Marine and Coastal Ecosystems in a Changing Arctic. Retrieved September 26, 2020, from https://oaarchive.arctic-council.org/handle/11374/413

Arctic Council. (2015c). PAME: Framework for a Pan-Arctic Network of Marine Protected Areas. Retrieved September 26, 2020, from https://oaarchive.arctic-council.org/handle/11374/417

Arctic Council. (2016). CAFF: Arctic biodiversity principles and their application in mainstreaming biodiversity. Phase 1: Background and options paper. Retrieved September 20, 2020, from https://www.caff.is/strategies-series/433-arctic-biodiversity-principles-and-their-application-in-mainstreaming-biodiversi

Arctic Council. (2017). Declaration of the Fairbanks Ministerial Meeting. Retrieved September 26, 2020, from https://oaarchive.arctic-council.org/handle/11374/1910

Arctic Council. (2019). Joint PAME, CAFF, AMAP, SDWG Ecosystem Approach Expert Group. Guidelines for Implementing an Ecosystem Approach to Management of Arctic Marine Ecosystems. Retrieved September 19, 2020, from https://oaarchive.arctic-council.org/handle/11374/2390

Arctic Council. (2020). PAME, The increase in Arctic Shipping 2013–2019. Arctic Sipping Status Report (ASSR), March 31, 2020. Retrieved December 18, 2020, from https://www.pame.is/document-library/shipping-documents/arctic-ship-traffic-data-documents/reports/752-arctic-shipping-report-1-the-increase-in-arctic-shipping-2013-2019-pdf-version-1/file

Arctic Ocean Conference. (2008). Ilulissat Declaration, 28 May 2008. Retrieved September 22, 2020, from https://cil.nus.edu.sg/wp-content/uploads/2017/07/2008-Ilulissat-Declaration.pdf

Barry, T., Daviðsdóttir, B., Einarsson, N., & Young, O. (2020a). The Arctic Council: An agent of change? *Global Environmental Change, 63*, 1–10.

Barry, T., Daviðsdóttir, B., Einarsson, N., & Young, O. (2020b). How does the Arctic Council support conservation of Arctic Biodiversity? *Sustainability, 12*, 1–18.

Bartenstein, K. (2011). The Arctic exception in the Law of the Sea Convention: A contribution to safer navigation in the Northwest Passage? *Ocean Development & International Law, 42*, 22–52.

Bird, K. J., Charpentier, R. R., Gautier, D. L., Houseknecht, D. W., Klett, T. R., Pitman, J. K., et al. (2008). *Circum-Arctic Resource Appraisal: Estimates of Undiscovered Oil and Gas North of the Arctic Circle*, U.S. Geological Survey. Retrieved September 17, 2020, from https://pubs.usgs.gov/fs/2008/3049/

Carson, M., (Dir.). (2016). *Arctic Resilence Report ARR*, Stockholm Environment Institute and the Stockholm Resilience Centre.

CBD. (1992). UNTS, vol, 1760, p. 79. Retrieved September 20, 2020, from https://www.cbd.int/convention/text/

CBD. (1998). Report of the Workshop on the Ecosystem Approach. Submission by the Governments of the Netherlands and Malawi. UNEP/CBD/COP/4/Inf.9. Retrieved September 20, 2020, from https://www.cbd.int/doc/meetings/cop/cop-04/information/cop-04-inf-09-en.pdf

CBD. (2000). Decision V/6: Ecosystem approach. Retrieved September 20, 2020, from https://www.cbd.int/decisions/cop/5/6

CBD. (2003). Report of the Ad Hoc Technical Expert Group on Marine and Coastal Protected Areas, UNEP/CBD/SBTTA/8/INF/7, 13 February 2003. Retrieved September 20, 2020, from https://www.cbd.int/kb/record/meetingDocument/4748?RecordType=meetingDocument&Event=SBSTTA-08

CBD. (2004a). Decision VII/5: Marine and coastal. Marine and coastal biological diversity. Retrieved September 20, 2020, from https://www.cbd.int/doc/decisions/cop-07/cop-07-dec-05-en.pdf

CBD. (2004b). Decision VII/11: Ecosystem approach. Retrieved September 20, 2020, from https://www.cbd.int/doc/decisions/cop-07/cop-07-dec-11-en.doc

CBD. (2004c). Secretariat: The ecosystem approach. Retrieved September 20, 2020, from https://www.cbd.int/doc/publications/ea-text-en.pdf

CBD. (2006). The Ecosystem Approach: Advanced user's guide. Retrieved September 20, 2020, from https://www.cbd.int/doc/programmes/cro-cut/eco/eco-guide-ad-en.pdf

CBD. (2010). Decision X/13: New and emerging issues. Retrieved September 20, 2020, from https://www.cbd.int/decision/cop/?id=12279

CBD. (2011). SBSTTA: Recommendation: *XV/7.* Arctic biodiversity. Retrieved September 20, 2020, from https://www.cbd.int/recommendation/sbstta/?id=12974

CBD. (2014). Report of the Arctic Regional Workshop to Facilitate the Description of Ecologically or Biologically Significant Marine Areas, UNEP/CBD/EBSA/WS/2014/1/5. Retrieved September 20, 2020, from https://www.cbd.int/doc/meetings/mar/ebsaws-2014-01/official/ebsaws-2014-01-05-en.pdf

CBD. (2015). Secretariat: EBSAS Regions Considered by the CBD Conference of the Parties. Retrieved September 20, 2020, from https://www.cbd.int/ebsa/ebsas?tab=considered

CBD. (2020). Preparation of the Post-2020 Global Biodiversity Framework. Draft recommendation submitted by the Co-Chairs, CBD/WG2020/2/L.2, 28 February 2020

Christiansen, J., Mecklenbirg, C., & Karamushko, O. (2014). Arctic marine fishes and their fisheries in light of global change. *Global Change Biology, 20*(2), 352–359. Retrieved August 27, 2020, from https://www.ncbi.nlm.nih.gov/pmc/articles/PMC4255237/

Comer, B., Osipova, L., Georgeff, E., & Mao, X. (2020). *The International Maritime Organization's Proposed Arctic Heavy Fuel Oil Ban: Likely impacts and opportunities for improvement.* The International Council on Clean Transportation, Washington, DC. Retrieved September 12, 2020, from https://theicct.org/publications/analysis-HFO-ban-IMO-2020

De Lucia, V. (2015). Competing narratives and complex genealogies: The ecosystem approach in international environmental law. *Journal of Environmental Law, 27*(1), 91–117.

De Lucia, V. (2017). The Arctic environment and the BBNJ negotiations. Special rules for special circumstances. *Marine Policy, 86*, 234–240.

De Lucia, V. (2019). The Ecosystem Approach and the negotiations towards a new Agreement on Marine Biodiversity in Areas beyond National Jurisdiction. *Nordic Environmental Law Journal, 2*, 7–25.

De Santo, E. M. (2018). Implementation challenges of area-based management tools (ABMTs) for biodiversity beyond national jurisdiction (BBNJ). *Marine Policy, 97*, 34–43.

Enright, S. R., & Boteler, B. (2020). The ecosystem approach in international marine law and governance. In T. O. O'Higgins, M. Lago, & T. H. DeWitt (Eds.), *Ecosystem-based management, Ecosystem services and aquatic biodiversity* (pp. 333–352). Switzerland: Springer Open.

French, N. (2017). Can the Ecosystem Approach (EA) work in Arctic science and governance?. *The Polar Connection.* Retrieved September 23, 2020, from http://polarconnection.org/ecosystem-approach/#:~:text=The%20EA%20in%20the%20Arctic,in%20order%20to%20reduce%20environmental

Gautier, D. L., Bird, K. J., Charpentier, R. R., Grantz, A., Houseknecht, D. W., Klett, T. R., et al. (2009). Assessment of undiscovered oil and gas in the Arctic. *Science, 324*, 1175–1179.

Gavrilov, V., Dremliuga, R., & Nurimbetov, R. (2019). Article 234 of the 1982 United Nations Convention on the law of the sea and reduction of ice cover in the Arctic Ocean. *Marine Policy, 106*, 1–6.

Hoel, A. A. (2010). Integrated oceans management in the Arctic: Norway and beyond. *Arctic Review on Law and Politics, 1*(2), 186–206.

Hossain, K., & Morris, K. (2017). Protecting Arctic Ocean Marine Biodiversity in the area beyond national jurisdiction. Plausible legal frameworks for protecting high Arctic waters. In G. Andreone (Ed.), *The future of the Law of the Sea. Bridging gaps between national, individual and common interests* (pp. 110–111). Springer Open.

ICES. (2018). Report of the Arctic Fisheries Working Group (AFWG), ICES CM 2018/ACOM:06. Retrieved September 26, 2020, from http://www.ices.dk/sites/pub/Publication%20Reports/Expert%20Group%20Report/acom/2018/AFWG/00-AFWG%202018%20Report.pdf

ICJ. (1996). Legality of the Threat or Use of Nuclear Weapons, Advisory Opinion, 1. C.J. Reports 1996, p. 226. Retrieved December 18, 2020, from https://www.icj-cij.org/public/files/case-related/95/095-19960708-ADV-01-00-EN.pdf

ICJ. (2010). Pulp Mills on the River Uruguay (Argentina v. Uruguay), Judgment, I.C.J. Reports 2010, p. 14, par. 184. Retrieved December 18, 2020, from https://www.icj-cij.org/en/case/135/judgments

IMO. (2010). Resolution MEPC.189(60). Amendments to the Annex of the Protocol of 1978 Relating to the International Convention for the Prevention of Pollution From Ships, 1973 (Addition of a new chapter 9 to MARPOL Annex I). Retrieved September 26, 2020, from http://www.imo.org/en/KnowledgeCentre/IndexofIMOResolutions/Marine-Environment-Protection-Committee-(MEPC)/Pages/MEPC-2010-11.aspx

IMO. (2015a). Resolution MEPC.264(68): International Code for Ships Operating in Polar Waters (Polar Code) and Resolution MEPC.265(68). Retrieved September 26, 2020, from http://www.imo.org/en/KnowledgeCentre/IndexofIMOResolutions/Marine-Environment-Protection-Committee-(MEPC)/Pages/MEPC-2014-15.aspx

IMO. (2015b). Polar Code, Part II-A, "Pollution Prevention Measures", MEPC 68/21/Add.1, Report of the Marine Environment Protection Committee on its sixty-eighth session, annex 10. Retrieved September 26, 2020, from https://www.register-iri.com/wp-content/uploads/MEPC_Resolution_26468.pdf

IMO. (2015c). MEPC: Impact on the Arctic of emissions of Black Carbon from international shipping, MEPC 68/21, 29 May 2015. Retrieved September 10, 2020, from http://www.imla.co/sites/default/files/mepc_68-21_-_report_of_the_marine_environment_protection_committee_on_its_sixty-eighth_session_secretariat.pdf

IMO. (2017a). MEPC: Protecting the Arctic from heavy fuel oil – work to begin at MEPC 72. Retrieved September 10, 2020, from http://www.imo.org/en/MediaCentre/MeetingSummaries/MEPC/Pages/MEPC-71.aspx

IMO. (2017b). Guide on Response to Oil Spills in Snow and Ice Conditions (Edition 2017)

IMO. (2018). List of Special Areas, Emission Control Areas and Particularly Sensitive Sea Areas, MEPC.1/Circ.778/Rev.3. Retrieved September 10, 2020, from http://www.gard.no/Content/26411326/IMO%20MEPC1-Circ778-Rev3_Special%20Areas%2C%20ECAs%20and%20PSSAs%20under%20MARPOL.pdf

IMO. (2019). Report of the Marine Environment Protection Committee on its seventy-fourth session: Impact on the Arctic of emissions of Black Carbon from international shipping, MEPC 74/18, 9 June 2019. Retrieved September 10, 2020, from https://www.iadc.org/wp-content/uploads/2019/08/MEPC-74-18-Report-Of-The-Marine-Environment-Protection-CommitteeOn-Its-Seventy-Fourth-Session-Secretariat.pdf

IMO. (2020a). Status of Conventions. Retrieved September 26, 2020, from http://www.imo.org/en/About/Conventions/StatusOfConventions/Pages/Default.aspxo

IMO. (2020b). Sub-Committee on Pollution Prevention and Response (PPR 7): Prohibiting the use and carriage for use as fuel of heavy fuel oil by ships in the Arctic waters- draft MARPOL amendments agreed. Retrieved September 26, 2020, from http://www.imo.org/en/MediaCentre/MeetingSummaries/PPR/Pages/PPR-7th-Session.aspx

IPCC. (2018). Global Warming of 1.5 °C. Retrieved April 29, 2021, from https://www.ipcc.ch/site/assets/up-load/sites/2/2019/06/SR15_Full_Report_High_Res.pdf

IPCC. (2019). IPCC Special Report on the Ocean and Cryosphere in a Changing Climate. Retrieved April 29, 2021, from https://www.ipcc.ch/srocc/download

ITLOS. (2015). Request for Advisory Opinion submitted by the Sub-Regional Fisheries Commission, Advisory Opinion, 2 April 2015, ITLOS Reports 2015, p. 4. Retrieved December 18, 2020, from https://www.itlos.org/fileadmin/itlos/documents/cases/case_no.21/advisory_opinion_published/2015_21-advop-E.pdf

Jakobsen, I. U. (2013). The adequacy of the law of the sea and international environmental law to the marine Arctic: Integrated ocean management and shipping. *Michigan State International Law Review, 22*, 291–320.

Jakobsen, I. U. (2016). *Marine protected areas in international law. An Arctic perspective*. Leiden, Boston: Brill Nijhoff.

Jentoft, S., & Bjorkan, M. (2007). Marine protected areas: A governance system analysis. *Human Ecology, 35*, 611–622.

Kirk, E. A. (2015). The ecosystem approach and the search for an objective and content for the concept of holistic ocean governance. *Ocean Development & International Law, 46*, 33–49.

Koester, V. (2018). Environmental principles and concepts in biodiversity treaties. In L. Krämer & E. Orlando (Eds.), *Principles of environmental law* (pp. 538–554). Cheltenham: Edward Elgar Publishing.

Koivurova, T. (2009). Governance of protected areas in the Arctic. *Utrecht Law Review, 5*(1), 44–60.

Koivurova, T., Molenaar, E., & VanderZwaag, D. (2009). Canada, the European Union, and Arctic Ocean Governance: A tangled and shifting seascape and future directions. In T. Koivurova, A. Chircop, E. Franck, E. J. Molenaar, & D. VanderZwaag (Eds.), *Understanding and strengthening European Union-Canada relations in law of the sea and ocean governance* (Vol. 35, pp. 107–146). Juridica Lapponica.

Lalonde, S. (2013). Marine protected areas in the Arctic. In E. Molenaar, A. G. Oude Elferink, & D. Rothwell (Eds.), *The law of the sea and the Polar Regions. Interactions between global and regional regimes* (pp. 85–111). Leiden-Boston: Martinus Nijhoff.

Langlet, D., & Rayfuse, R. (2019). *The ecosystem approach in ocean planning and governance: Perspectives from Europe and Beyond*. Leiden; Boston: Brill Nijhoff.

Larose, C., Dommerque, A., & Vogel, T. (2013). The dynamic Arctic snow pack: An unexplored environment for microbial diversity and activity. *Biology, 2*(1), 317–330.

Malinauskaite, L., Cook, D., Brynhildur, D., Ögmundardóttir, H., & Roman, J. (2019). Ecosystem services in the Arctic: A thematic review. *Ecosystem Services, 36*, 1–14.

MARPOL Convention. (1973/1978). UNTS, vol. 1340, p. 61. Retrieved September 6, 2020, from https://treaties.un.org/doc/Publication/UNTS/Volume%201340/volume-1340-A-22484-English.pdf

McDonough, K., Hutchinson, S., Moore, T., & Hutchinson, J. M. S. (2017). Analysis of publication trends in ecosystem services research. *Ecosystem Services, 25*, 82–88.

McRae, D. M., & Goundrey, D. J. (1982). Environmental jurisdiction in Arctic waters: The extent of Article 234. *University of British Columbia Law Review, 126*, 211–228.

Molenaar, E. (2009). Arctic marine shipping: Overview of the international legal framework, gaps, and options. *Journal of Transnational Law and Policy, 18*, 288–325.

Molenaar, E., & Oude Elferink, A. (2009). Marine protected areas in areas beyond national jurisdiction-the pioneering efforts under the OSPAR Convention. *Utrecht Law Review, 1*, 5–20.

Osofsky, H., Shadian, J., & Fechtelkotter, S. (2016). Arctic Energy Cooperation. U.C. *Davis Law Review, 49*, 1431–1510.

OSPAR. (2003). OSPAR Commission: Towards an Ecosystem Approach to the Management of Human Activities. Retrieved September 26, 2020, from https://www.ospar.org/site/assets/files/1232/jmm_annex05_ecosystem_approach_statement.pdf

OSPAR. (2010a). Recommendation 2003/3, amended by Recommendation 2010/2, Annex 7. Retrieved September 26, 2020, from https://www.ospar.org/search?q=Recommendation+2010%2F2#results

OSPAR. (2010b). The North-East Atlantic Environment Strategy Strategy of the OSPAR Commission 2010–2020 (OSPAR Agreement 2010-3). Retrieved September 26, 2020, from https://www.ospar.org/site/assets/files/1200/ospar_strategy.pdf

OSPAR Convention. (1992). UNTS, vol. 2354, p. 67. Retrieved September 26, 2020, from https://www.ospar.org/convention/text

OSPAR MPA database: Map Tool. Retrieved September 10, 2020., from http://carto.mpa.ospar.org/1/ospar.map

PCA. (2016). Permanent Court of Arbitration, Case 2013-19, In the Matter of the South China Arbitration (The Republic of Philippines v. The People's Republic of China), 12 July 2016. Retrieved December 12, 2020, from https://pca-cpa.org/en/cases/7/

Polakowski, H. (2017). Freezing the issues: Why Arctic coastal states need to implement marine protected areas in the arctic seas. *Tulane Environmental Law Journal, 30*(2), 347–368.

Rayfuse, R. (2014). Coastal state jurisdiction and the Polar Code: A test for Arctic Oceans governance? In T. Stephens & D. L. VanderZwaag (Eds.), *Polar ocean governance in an era of environmental change* (pp. 235–252). Edward Elgar Publishing.

Rotten, S. V. (2020). *The Arctic Council. Between environmental protection and geopolitics.* Springer.

Sebek, V. (1990). The North Sea and the concept of special areas. *International Journal of Estuarine and Coastal Law, 5*, 157–166.

Smith, D., & Jabour, J. (2018). MPAs in ABNJ: Lessons from two high seas regimes. *ICES Journal of Marine Science, 75*(1), 417–425.

Trouwborst, A. (2009). The precautionary principle and the ecosystem approach in international law: Differences, similarities and linkages. *Review of European, Comparative & International Environmental Law, 18*(1), 26–37.

UNCLOS. (1982). UNTS vols. 1833-1834-1835. Retrieved February 8, 2020, from https://treaties.un.org/doc/Publication/UNTS/Volume%201833/volume-1833-A-31363-English.pdf

UNGA. (2015). Res. A/69/292: Development of an international legally binding instrument under the United Nations Convention on the Law of the Sea on the conservation and sustainable use of marine biological diversity of areas beyond national jurisdiction. Retrieved April 10, 2020, from https://sustainabledevelopment.un.org/index.php?page=view&type=111&nr=7897&menu=35

Van Dyke, J., & Broder, S. (2012). Particularly sensitive sea areas. Protecting the marine environment in the territorial seas and exclusive economic zones. *Denver Journal of International Law & Policy, 40*, 472–481.

Waylen, K., Blackstock, K., & Holstead, K. (2013). *Exploring experiences of the Ecosystem Approach, November 2013.* The James Hurron Institute.

Waylen, K., Hastings, E., Banks, E., Holstead, K., Irvine, R., & Blackstock, K. (2014). The need to distangle key concepts from ecosystem-approach jargon. *Conservation Biology, 28*(5), 1215–1224.

Young, O. (2016). Governing the Arctic Ocean. *Marine Policy, 72*, 271–277.

Chapter 10
Governance of the Caribbean Marine Biodiversity: Complex Challenges for a Complex Region

Luis E. Rodríguez-Rivera

Contents

10.1 Introduction 180
10.2 Marine Biodiversity in the Caribbean Region 181
10.3 The Caribbean Sea: Biodiversity Hotspot in Peril 181
10.3.1 Local/Direct Factors 183
10.3.2 Local/Indirect Factors 184
10.3.3 External/Direct Factors 184
10.3.4 External/Indirect Factors 185
10.4 Marine Biodiversity Management Approaches in Caribbean Region 186
10.4.1 Caribbean Large Marine Ecosystem (CLME) 186
10.4.2 Caribbean Large Marine Ecosystem and Adjacent Areas (CLME+) 186
10.4.3 Caribbean Islands Biodiversity Hotspot (CIBH) 187
10.4.4 Wider Caribbean Region (WCR) 187
10.4.5 Greater Caribbean Region (GCR) 188
10.5 Governance Structure of Marine Biodiversity Management in the Caribbean Region 188
10.5.1 International Institutional Frameworks Pertinent to Marine Biodiversity in the Caribbean Region 189
10.5.2 Regional Institutional Frameworks Pertinent to Marine Biodiversity in the Caribbean Region 190
10.5.3 International Legal Frameworks: Multilateral Environmental Agreements Pertinent to Marine Biodiversity in the Caribbean Region 193
10.5.4 Regional Legal Frameworks: Multilateral Agreements Pertinent to Marine Biodiversity in the Caribbean Region 195
10.5.5 Non-Governmental Organizations (NGOs) 196
10.6 Challenges for Effective Marine Biodiversity Governance in the Caribbean Region .. 196
10.7 Conclusion 198
References 198

Abstract This chapter discusses the state of biodiversity in the Caribbean Sea: rich and vibrant, yet threatened and vulnerable. It also examines governance of marine biodiversity in the Caribbean Region, which is based on a structure made up of an

---

L. E. Rodríguez-Rivera (✉)
Universidad de Puerto Rico, Law School, San Juan, PR, USA
e-mail: luis.rodriguez89@upr.edu

M. Campins Eritja, T. Fajardo del Castillo (eds.), *Biological Diversity and International Law*, https://doi.org/10.1007/978-3-030-72961-5_10

intricate, overlapping web of international and regional institutions; multilateral biodiversity, environmental and ocean management agreements; and international and regional policies, plans and initiatives implemented at the national and local levels. The effectiveness of the governance framework for Caribbean marine biodiversity is weighed against the most prevalent political, socio-economic and climate challenges facing the region. Lastly, the chapter sets out recommendations for a future agenda towards the development of a more effective governance structure for the Caribbean's marine biodiversity.

10.1 Introduction

As the thirtieth anniversary of the Convention on Biological Diversity fast approaches, Caribbean countries, in particular the Caribbean islands, face unprecedented challenges as they recover from devastating tropical storms and climate events that have destroyed their essential infrastructure and economies, and further depleted the region's fragile and vulnerable marine biodiversity ecosystems, habitats and species. Caribbean countries participate in a host of multilayered international and regional biodiversity, environmental and ocean management frameworks, agreements, organizations and initiatives aimed at the conservation, protection and sustainable management of marine resources and biodiversity in the Caribbean Sea. However, the Caribbean Sea's once rich and vibrant biodiversity has steadily deteriorated, while becoming more vulnerable to increasing human and natural threats. The growing deterioration and vulnerability of Caribbean biodiversity, in turn, threatens the very livelihood of Caribbean people whose dependence on the region's marine biodiversity is directly linked to their countries' food security, economic development and tropical storm protection, especially in the case of the Caribbean islands.

After reviewing the state of biodiversity in the Caribbean Sea and the key factors that have led to its present vulnerable and threatened condition, this chapter examines the regional marine ecosystem management approaches implemented in the Caribbean region, and the governance structure in place for the development and implementation of the same, including the international and regional institutional and legal frameworks used to implement them. Particular attention is placed on the Caribbean Islands Biodiversity Hotspot as it encloses the Caribbean Sea's most valuable biodiversity resources. Further, the chapter considers the environmental, social, economic and political complexities found in the Caribbean region that have contributed to the deterioration of its marine biodiversity as well as inherent flaws found in the region's fragmented and uncoordinated marine biodiversity governance structure.

The conservation and sustainable management of the Caribbean Sea's biodiversity resources involve complex multisectoral interactions, arrangements and policies. A new, broader agenda is proposed in this chapter, and minimum requirements are recommended.

10.2 Marine Biodiversity in the Caribbean Region

The Caribbean Sea, a semi-enclosed body of water in the western part of the Atlantic Ocean, is the second largest sea in the world. Bounded to the north by the Greater Antilles Islands, to the east by the Lesser Antilles Islands, to the south by the northern coasts of South America and to the west by the eastern coasts of Central America, the Caribbean Sea is the defining element of the region commonly called the "Caribbean" (Parris 2016; Agard and Cropper 2007).

The Caribbean Sea holds well over 12,000 marine species, which represent the greatest concentration of such species in the entire Atlantic Ocean (CEPF 2019; Miloslavich et al. 2010). The region's most valuable marine biodiversity is contained within the Caribbean Islands Biodiversity Hotspot (CIBH). The CIBH consists of approximately "7,000 islands, islets, reefs and cays with a land area of 230,000 km^2 scattered across 4 million km^2 of sea" (CEPF 2019, p. 21). Thirty sovereign states and non-sovereign territories are located within the CIBH (Table 10.1).

The CIBH's shallow marine environment and coastal zones sustain productive and biologically complex ecosystems, such as coral reefs, seagrass beds, soft sea bottoms, coastal wetlands, mangroves and beaches (Robertson and Cramer 2014; Miloslavich et al. 2010). The ecosystem services provided by these habitats are invaluable to the Caribbean countries' livelihood as they provide habitat to numerous species, support fisheries and tourism, and protect the coast from extreme climate events (CEPF 2019; Nieto-Blázquez et al. 2017; Spalding et al. 2001).

10.3 The Caribbean Sea: Biodiversity Hotspot in Peril

During the past five decades, the rich and valuable biodiversity resources within the Caribbean Sea's marine environment have dramatically deteriorated. For instance, since 1980 the area coverage of mangroves has decreased by approximately 1% per year; live coral cover has declined up to 80% in many areas; and approximately 35% of the region's fish stocks are overexploited (Miloslavich et al. 2010).

The key factors or drivers of change responsible for the depletion of the Caribbean's marine biodiversity are succinctly outlined and classified in the 2007 Caribbean Sea Ecosystem Assessment (Agard and Cropper 2007), which sets out four categories: local and direct, local and indirect, external and direct, and external and indirect. These drivers have particularly impacted the marine biodiversity within the CIBH. Moreover, said drivers are still prevalent today, and have indeed multiplied and intensified, as outlined and updated below (see for example: CEPF 2019;

Table 10.1 List of countries contained within Caribbean regions and organizations (* indicates associate members) (Table prepared by author)

	CLME	CLME +	CIBH	WCR	GCR	ACS	CARICOM	OECS
North America								
United States	x	x		x	x			
Central America								
Belize	x	x		x	x	x	x	
Costa Rica	x	x		x	x	x		
El Salvador						x		
Guatemala	x	x		x	x	x		
Honduras	x	x		x	x	x		
Mexico	x	x		x	x	x		
Nicaragua	x	x		x	x	x		
Panama	x	x		x	x	x		
South America								
Brazil		x			x			
Colombia	x	x		x	x	x		
Guyana		x		x	x	x	x	
Suriname		x		x	x	x	x	
Venezuela	x	x		x	x	x		
French Guiana (France)		x		x	x	x*		
Islands Sovereign								
Antigua and Barbuda	x	x	x	x	x	x	x	x
Bahamas	x	x	x	x	x	x	x	
Barbados	x	x	x	x	x	x	x	
Cuba	x	x	x	x	x	x		
Dominica	x	x	x	x	x	x	x	x
Dominican Republic	x	x	x	x	x	x		
Grenada	x	x	x	x	x	x	x	x
Haiti	x	x	x	x	x	x	x	
Jamaica	x	x	x	x	x	x	x	
Saint Kitts and Nevis	x	x	x	x	x	x	x	x
Saint Lucia	x	x	x	x	x	x	x	x
St. Vincent & the Grena.	x	x	x	x	x	x	x	x
Trinidad and Tobago	x	x		x	x	x	x	
Islands Non-Sovereign								
Anguilla (UK)	x	x	x	x	x		x*	x*
Aruba (Netherlands)	x	x	x	x	x	x*		
Bermuda (UK)							x*	
Bonaire (Netherlands)	x	x	x	x	x	x*		

(continued)

Table 10.1 (continued)

	CLME	CLME +	CIBH	WCR	GCR	ACS	CARICOM	OECS
British Virgin Is. (UK)	x	x	x	x	x	x*	x*	x*
Cayman Is. (UK)	x	x	x	x	x		x*	
Curaçao (Netherlands)	x	x	x	x	x	x*		
Guadeloupe (France)	x	x	x	x	x	x*		x*
Martinique (France)	x	x	x	x	x	x*		x*
Montserrat (UK)	x	x	x	x	x		x	x
Navassa (US)			x					
Puerto Rico (US)	x	x	x	x	x			
Saba (Netherlands)	x	x	x	x	x	x*		
Saint Barthélemy (Fra.)	x	x	x	x	x	x*		
Saint Eustatius (Nether.)	x	x	x	x	x	x*		
Saint Martin (France)	x	x	x	x	x	x*		
Saint Maarten (Nether.)	x	x	x	x	x	x*		
Turks & Caicos Is. (UK)	x	x	x	x	x		x*	
US Virgin Islands (US)	x	x	x	x	x			

CARICOM 2018; UNEP–WCMC 2016; Pittman et al. 2015; Miloslavich et al. 2010; Fanning et al. 2009; Singh 2008).

10.3.1 Local/Direct Factors

- Changes in coastal land use and sea use associated with urbanization, industry, commerce, ports and harbours and tourism-related developments. Result: depletion and destruction of marine biodiversity habitats (i.e. coral reefs, seagrass beds, mangroves and beaches) and the ecosystem services they provide.
- Contamination of coastal environments from land-based point and nonpoint sources, deforestation and agricultural water runoff. Result: increase in nutrient levels leading to elevated algae growth and the introduction of sediments and chemicals which, in turn, deteriorate and deplete marine biodiversity habitats and reduce their biodiversity ecosystem services.
- Contamination of marine environments from offshore oil and gas extraction, mining and maritime transport (cargo trans-shipment and cruise ships). Result:

destruction of marine biodiversity habitats (i.e. coral reefs, seagrass beds, mangroves and beaches) and ecosystem services.
- Overexploitation of coastal and offshore living marine resources, such as overfishing and use of inadequate fishing practices at both the commercial and subsistence levels as well as illegal, unregulated and unreported fishing. Result: deterioration of marine biodiversity habitats (i.e. coral reefs and seagrass beds) and reduction in fish and other living marine populations.

10.3.2 Local/Indirect Factors

- Dependency on non-sustainable tourism as the principal economic development strategy. Result: overdevelopment of coastal areas which, in turn, deteriorates biological diversity and reduces the coastal ecosystem's capacity to provide services.
- Lack of effective and coordinated regional governance for the shared marine living resources in the Caribbean Sea. Result: overexploitation of shared living marine resources which, in turn, deteriorates and destroys ecosystems and reduces biodiversity.
- Lack of governance capacities (i.e. public administration, human resources, and technical/scientific, financial and monitoring capacities) needed for effective biodiversity conservation and management at national and regional levels. Result: ineffective policies, programs and initiatives at the national level, which, in turn, result in deterioration and destruction of coastal and marine biodiversity ecosystems and services.
- Population growth, increased consumption, poverty and inequality. Result: increased pressure on unsustainable land-use practices which, in turn, have an impact on the capacity of marine biodiversity ecosystems and ecosystem services.

10.3.3 External/Direct Factors

- Rising sea temperatures and increasing intensity and frequency of tropical storm events related to climate change. Result: destruction of coral reefs, beaches and other marine habitats, and of the biological diversity and services they produce.
- The islands of the Caribbean region are, literally and figuratively, in the eye of the storm. Commonly referred to as Hurricane Alley, the Caribbean region is threatened yearly by seasonal tropical storms and hurricanes. In the period between 1851 and 2019, at least one tropical storm developed into a major hurricane in 137 of the 169 hurricane seasons (López-Marrero and Castro-Rivera 2019; Castro-Rivera and López-Marrero 2019). More recently, the frequency and intensity of tropical storms have intensified to alarming levels. In 2017 and

2019, the Caribbean islands within the CIBH experienced extraordinary hurricane seasons in terms of both meteorological standards and socio-economic impacts:

> [Hurricane] Irma hit four island groupings as a category 5 hurricane: Barbuda, Sint Maarten/Saint Martin, the British Virgin Islands, and Cuba. Hurricane Maria struck Dominica as [a] category 5 and Puerto Rico as [a] category 4 (very close to category 5). Rarely have two category 4 and 5 hurricanes made landfall in the region in the same year. ... At the time, Hurricane Irma was the strongest hurricane in the Caribbean Sea on record. ... In addition, Irma was the strongest hurricane to hit the Leeward Islands. The 2017 season was also characterized by major back-to-back hurricanes in the region: Irma, Jose and Maria. Two years later, during the 2019 season, another category-5 hurricane, Dorian, directly impacted the Bahamas ... where it remained stationary for about 24 hours. Dorian reached winds of 298 km/h, thus matching Irma's record, but it possessed a lower atmospheric pressure. ... Hurricanes of these two seasons caused massive destruction and devastation, loss of life and property, disruption to daily lives and well-being, major economic damages in cases setting back years of economic development, and unprecedented human displacements in the Caribbean region (Castro-Rivera and López-Marrero 2019, pp. 27–28).

- The US National Oceanic and Atmospheric Administration (NOAA) predicted "above-normal activity" for the 2020 hurricane season owing to multiple factors relating to climate, including "warmer-than-average sea surface temperatures in the tropical Atlantic Ocean and Caribbean Sea" (US NOAA 2020).
- The destruction caused by stronger and more frequent tropical storms significantly exacerbates the deterioration of coral reefs, mangroves, beaches and other important marine biodiversity habitats in the Caribbean region. For example, studies on the effects of the 2017 hurricanes Irma and Maria on Puerto Rico's marine biodiversity habitats reveal: 11% of the island's coral reefs were damaged (US NOAA 2018); diverse and complex geomorphic changes occurred to the island's beach systems (Barreto-Orta et al. 2019); and there was a 53% mortality rate for the island's mangroves and coastal wetlands (Branoff et al. 2018).
- Rising sea-levels caused by climate change. Result: saltwater intrusion into the coastal environment which, in turn, leads to displacement and destruction of mangroves and coastal wetlands (Yu et al. 2019).
- Interaction between the Caribbean Sea and effluent discharges from the Magdalena, Orinoco and Amazon rivers. Result: deterioration of marine biodiversity ecosystems and species because of sediment and toxic contamination.
- Introduction of invasive species through maritime transport and tourism activities (e.g., lionfish). Result: destruction of marine biodiversity habitats and species.

10.3.4 External/Indirect Factors

- The presence of the Panama Canal and the implementation of innocent, transit and archipelagic passage through the Caribbean Sea, pursuant to the United Nations Convention on the Law of the Sea (UNCLOS), produce excessive and concentrated maritime traffic through the Caribbean Sea. Result: increased risk of significant toxic and radioactive contamination as well as increased vulnerability of the region's marine biodiversity habitats and species (Rodríguez-Rivera 2014).

10.4 Marine Biodiversity Management Approaches in Caribbean Region

Several marine ecosystem conservation and management approaches have been implemented in the Caribbean region, including among others: the Caribbean Large Marine Ecosystem, the Caribbean Large Marine Ecosystem and Adjacent Areas, the Caribbean Islands Biodiversity Hotspot, the Wider Caribbean Region, and the Greater Caribbean Region. As identified below, a total of 44 sovereign states and non-sovereign territories located within, around or adjacent to the Caribbean Sea influence the areas defined by said regional marine ecosystem conservation and management approaches.

10.4.1 Caribbean Large Marine Ecosystem (CLME)

Large marine ecosystems (LMEs) are "extensive areas of ocean space of $\geq$ 200,000 km^2, characterized by distinct hydrographic regimes, submarine topography, productivity, and trophically dependent populations" (Sherman 1991, p. 349). The LME research and management approach for the conservation and management of living marine resources is consistent with the "ecological concept that critical processes controlling the structure and function of biological communities can best be addressed on a regional basis" (Sherman 1991, p. 349), and it finds its legal foundation in the provisions of UNCLOS that address the conservation and management of marine living resources (UNCLOS 1982).

The CLME is one of 66 LMEs in the world (LMEH 2020). The CLME is interconnected with three other LMEs: the Gulf of Mexico LME (GOM LME) to the northwest, the North Brazil Shelf LME (NBS LME) to the southwest, and the Southeast US Continental Shelf LME (SEUS LME) to the north. Located within or surrounding the CLME are 23 sovereign states and 17 non-sovereign territories (Table 10.1).

10.4.2 Caribbean Large Marine Ecosystem and Adjacent Areas (CLME+)

The CLME+ region refers to an ecosystem research and management approach for the living marine resources contained within the CLME and the NBS LME areas.

The oceanography of the Caribbean Sea is influenced by the interaction of currents from both neighbouring LMEs (Fanning et al. 2009). Contained within the CLME+ region are 26 sovereign states and 18 non-sovereign territories, including all 40 CLME countries as well as the sovereign states of Guyana, Suriname and Brazil and the non-sovereign territory of French Guiana (Mahon et al. 2013) (Table 10.1). As will be discussed below, the CLME+ region approach has been incorporated into the CLME+ Project developed by the United Nations Development Programme (UNDP) and the Global Environment Facility (GEF).

10.4.3 Caribbean Islands Biodiversity Hotspot (CIBH)

The designation of biodiversity hotspot is assigned to regions that meet two criteria: they have at least 1500 endemic species of vascular plants and they retain 30% or less of their original vegetation (Mittermeier et al. 2004; Meyers et al. 2000). The CIBH, which is one of 36 regions to have received this designation, is considered one of the most important centres of endemic biodiversity in the world (CEPF 2019; Nieto-Blázquez et al. 2017; Miloslavich et al. 2010; Lausche 2008). As defined, the CIBH region includes 29 of the 30 sovereign and non-sovereign islands and archipelagos located within the CLME (Trinidad and Tobago is excluded from the CIBH) in addition to the uninhabited US territory of Navassa (Table 10.1).

10.4.4 Wider Caribbean Region (WCR)

The 1983 Convention for the Protection and Development of the Marine Environment of the Wider Caribbean (known as the Cartagena Convention) defines the WCR as "the marine environment of the Gulf of Mexico, the Caribbean Sea and the areas of the Atlantic Ocean adjacent thereto" (Cartagena Convention 1983, Art. 2.1). The convention area includes the GOM LME, the CLME and the northern part of the NBS LME. The rationale for excluding the southern portion of the NBS LME corresponds more to political and geographical criteria than to an ecosystem management one as the convention's membership does not extend past the coast of French Guiana (France) on the northeast coast of South America. Hence, 43 of the 44 sovereign and non-sovereign countries located within the three interacting LMEs above are contained within the WCR (excluding Brazil and its portion of the NBS LME) (Table 10.1). In 1976, the Regional Seas Programme (RSP) of the United Nations Environment Programme (UNEP) established the Caribbean Environment Programme (UNEP CEP), which applies the Cartagena Convention as its legal framework (UNEP 2020d). Hence, the UNEP RSP has also incorporated the Cartagena Convention's definition of the WCR.

10.4.5 Greater Caribbean Region (GCR)

Conceptualized as a biogeographic region, the GCR includes areas contained within four LMEs: the GOM LME, the CLME, the NBS LME and the SEUS LME (Robertson and Cramer 2014). Although it represents a larger area than the WCR from an ecosystem management perspective, the GCR includes only one more country than the WCR: Brazil.

As further explained below, the Association of Caribbean States has also used the term "Greater Caribbean" to describe the scope of its sustainable development agenda among the states and territories of the Caribbean. However, the marine area of the ACS membership is smaller than the biogeographic GCR zone as it excludes the SEUS LME and the Brazilian portion of the NBS LME. Neither the US nor Brazil is a member of the ACS, which may explain this discrepancy, but the confusion remains.

10.5 Governance Structure of Marine Biodiversity Management in the Caribbean Region

The Caribbean region is one of the most politically, economically, socially and culturally complex and diverse regions in the world (Fanning et al. 2009, 2013; Lausche 2008; Agard and Cropper 2007). With 24 sovereign and non-sovereign islands on the United Nation's list of Small Island Developing States (SIDS) (UNOHRLLS 2020), the Caribbean Sea encloses the largest concentration of SIDS found in any of the world's LMEs. The United Nations General Assembly (UNGA) has recognized that the Caribbean SIDS "are ecologically fragile and socially and economically vulnerable and are also affected, inter alia, by their limited capacity, narrow resource base, need for financial resources, high levels of poverty and the resulting social problems, and the challenges and opportunities of globalization and trade liberalization" (UNGA 2016). These conditions, in turn, aggravate the perilous state of the region's marine biodiversity as

> [u]nsustainable exploitation of living resources, habitat degradation and destruction, and pollution of the marine environment are interlinked [in the Caribbean Sea], not only because of their synergistic impacts on living marine resources, but also because they have the same underlying socio-economic, legal and political root causes (Fanning et al. 2009, p. 226; Scobie 2012).

The synergistic impacts on the Caribbean's marine biodiversity are further aggravated by the fact that the region is extremely vulnerable to natural disasters related to climate change. As explained above, the Caribbean region has experienced an increase in the occurrence and intensity of tropical storms and hurricanes during the past decade, while rain patterns have altered dramatically, resulting in unseasonal floods and droughts. In turn, these climate change events threaten human life, health and safety (i.e., water availability, food security and infrastructure) as well as the region's biodiversity.

The "many transboundary oceanographic and ecological linkages within the Wider Caribbean" require a regional governance framework for the conservation, protection and management of the Caribbean region's interconnected marine biodiversity resources (Chakalall et al. 2007, p. 93). However, a regional ecosystem management approach is made difficult by the high fragmentation of marine areas within the Caribbean Sea resulting from the jurisdictional maritime zones codified in UNCLOS.

In response to the complex reality present in the Caribbean, international, regional, national and local efforts have coalesced towards the implementation of collaborative regional ecosystem management approaches for the shared marine resources in the Caribbean Sea. These efforts have driven Caribbean countries to participate in a host of multilayered international and regional biodiversity, environmental and ocean management frameworks, agreements, organizations and initiatives aimed at the conservation, protection and sustainable management of marine resources and biodiversity in the Caribbean Sea. Taken together, these institutional and legal frameworks constitute the region's governance structure for the conservation, protection and sustainable management of the Caribbean Sea's marine biodiversity. Below is an overview of the most important elements of this governance structure, with particular focus on those applicable to the CIBH region which contains the Caribbean region's most valuable marine biodiversity resources.

10.5.1 International Institutional Frameworks Pertinent to Marine Biodiversity in the Caribbean Region

10.5.1.1 United Nations Environment Programme (UNEP) Regional Seas Programme (RSP)

The conservation and protection of the Caribbean Sea, together with the region's need for sustainable development and improved management of its shared marine resources, have been a concern of the international community for nearly five decades. At the urging of several Caribbean countries, the UNEP Governing Council instituted the RSP in 1974 to address the accelerated degradation of the world's marine and coastal ecosystems (UNEP 1981). Through an integrated, cooperative and regional approach, countries around the world were encouraged to actively participate in the management and protection of their common marine environments (UNEP 2020b, c). At the time, UNEP's governing body also emphasized "the importance of activities in the Caribbean" (UNEP 1981). In 1976, the UNEP RSP launched the UNEP CEP as an endeavour aimed at facilitating the regional management of the shared marine resources within the Caribbean Sea (UNEP 2020d). The UNEP CEP is responsible for monitoring compliance with protocols, promoting regional environmental education, and coordinating data and information between members.

In 1977, the UNEP RSP and the Economic Commission for Latin America (ECLA) joined efforts to assess the quality of the Caribbean Sea's marine ecosystem and the causes leading to its deterioration. These efforts led to the 1981 Intergovernmental Meeting on the Action Plan for the Caribbean Environment Programme, where 22 Caribbean countries approved the Caribbean Action Plan (CAP) for the management and protection of the Caribbean Sea (UNEP 1981).

In 1983, the Cartagena Convention was adopted and became the legal framework for UNEP CEP's work in support of the regional management of marine resources within the WCR (UNEP 2012a). In 1986, the Regional Coordinating Unit (UNEP CAR/RCU) was established to administer the Cartagena Convention and its protocols, including the CAP (UNEP 2020a). Since then, the UNGA has recognized the Caribbean Sea on ten occasions as a "special area in the context of sustainable development," while reaffirming the region's unique biodiversity, its highly fragile ecosystem and its significant social, economic, environmental and climatological vulnerabilities (UNGA 2016).

10.5.1.2 United Nations Development Programme (UNDP) and Global Environment Facility (GEF) CLME+ Project

The CLME+ Project is a five-year project (2015–2020) implemented by UNDP, co-funded by GEF and executed by the United Nations Office of Project Services on behalf of the CLME+ countries (UNDP and GEF 2020) (Table 10.1). Through implementation of the Strategic Action Programme for the Sustainable Management of the Shared Living Marine Resources of the Caribbean and North Brazil Shelf Large Marine Ecosystems (CLME+ SAP), the CLME+ Project aims to assist the region in the research and management of its shared living marine resources (UNDP and GEF 2013). The CLME+ SAP is designed to facilitate interaction among a wide range of local, regional and international stakeholders and provide financial and technical support for the development and integration of the sustainability initiatives in the region. The UNGA has recognized the CLME+ Project's significant contribution to ocean governance in the Caribbean region (UNGA 2016).

10.5.2 Regional Institutional Frameworks Pertinent to Marine Biodiversity in the Caribbean Region

10.5.2.1 Association of Caribbean States (ACS)

Established in 1994, the ACS is a regional organization composed of 37 members and associate members with an agenda aimed at promoting cooperation in the Caribbean region. The ACS's work is directed primordially to 5 areas of concern:

preservation and conservation of the Caribbean Sea; sustainable tourism; trade and economic external relations; natural disasters; and transport (ACS n.d.). As noted above, the ACS incorporated the "Greater Caribbean" concept as its political framework. However, this "Greater Caribbean" concept links only 36 of the 43 sovereign states and non-sovereign territories located within the biogeographic GCR ecosystem management approach (Table 10.1). Furthermore, only 23 of the 30 sovereign and non-sovereign islands within the CIBH are part of the ASC (Table 10.1). The "Greater Caribbean" political concept also adds El Salvador, which is not included in any of the existing marine ecosystem management approaches as the country's coast lies on the Pacific Ocean side of Central America (ACS 2020b) (Table 10.1). Nonetheless, the ACS has played a leading role in the development of conservation and management policies in the Caribbean region.

The UNGA has recognized the ASC's development and implementation of regional initiatives aimed at the sustainable conservation and management of coastal and marine resources as well as the association's work towards the international recognition of the Caribbean Sea as a "special area in the context of sustainable development" (also referred to as the Caribbean Initiative) (UNGA 2016). In 2006, the ACS created the Caribbean Sea Commission (CSC), which coordinates and supports initiatives that promote the sustainable development of the Caribbean region (ACS 2020a). Although marine biodiversity conservation, protection and management in the Caribbean Sea are not the central focus of the ACS or the CSC, their efforts on natural disaster preparedness and sustainable tourism are pertinent and important (CEPF 2019).

10.5.2.2 Caribbean Community (CARICOM)

CARICOM is a regional intergovernmental organization that promotes and supports the economic integration of the Caribbean region. CARICOM is committed, among other things, to the sustainable management and conservation of its members' shared marine biodiversity (CARICOM 2001, Art. 58). Its membership and associate membership include 3 continental sovereign states, 11 sovereign island states and 6 non-sovereign island territories (CARICOM 2020) (Table 10.1). CARICOM's shared marine resources are located within the CLME and the northern portion of the NBS LME (except for Bermuda's marine ecosystem within the Atlantic Ocean, which lies outside of the CLME and the three LMEs that surround it).

In 2003, CARICOM adopted an agreement to establish the Caribbean Community Common Fisheries Policy (CCCFP), which entails a binding treaty for the conservation and sustainable management of fisheries and related ecosystems (CARICOM–CCCFP). Of the 30 sovereign and non-sovereign islands within the CIBH region, 15 are members or associate members of CARICOM (Table 10.1).

CARICOM's Secretariat serves as the Caribbean Hub of the Programme for Capacity Building Related to Multilateral Environmental Agreements (MEAs) in African, Caribbean and Pacific Countries (ACP), which is financed by the European Union and coordinated by UNEP. This programme aims to strengthen the capacity of Caribbean countries to implement MEAs (including those related with the conservation, protection and management of marine biodiversity) by providing technical and policy support as well as training. In addition, the ACP MEA programme has supported CARICOM's preparation of an assessment published under the title of *The State of Biodiversity in the Caribbean Community: A Review of Progress Towards the Aichi Biodiversity Targets* (CARICOM 2018).

10.5.2.3 Organisation of Eastern Caribbean States (OECS)

The OECS focuses on the regional integration of the Eastern Caribbean region (Lesser Antilles). Its membership includes 6 sovereign island states and 5 non-sovereign island territories that are all within the CLME and the CIBH areas (OECS 2010) (Table 10.1). The protection and conservation of its members' shared marine biodiversity is one of the OECS's guiding principles (OECS 2006, Prin. 13). The 2001 St. George's Declaration of Principles for Environmental Sustainability in the OECS, which was revised in 2006 (OECS 2006), and the 2002 OECS Environmental Management Strategy (OECS 2016) provide the region's framework for marine biodiversity conservation and sustainable management.

10.5.2.4 Caribbean Regional Fisheries Mechanism (CRFM)

The CRFM is the region's intergovernmental fisheries organization. In 2010, the CRFM adopted the Castries (Saint Lucia) Declaration on Illegal, Unreported and Unregulated Fishing, and it approved CARICOM's Common Fisheries Policy (CCCFP) in 2011. Much of CRFM's work is pertinent to the protection of marine biodiversity habitats and species within the CIBH region.

10.5.2.5 Sub-Commission for the Caribbean and Adjacent Regions of the Intergovernmental Oceanographic Commission for UNESCO (IOCARIBE)

IOCARIBE provides the Caribbean countries with marine scientific research and other related support, including training and education on biodiversity conservation and management. This technical and scientific support represents a significant contribution to many countries within the CIBH region that lack these capacities.

10.5.3 International Legal Frameworks: Multilateral Environmental Agreements Pertinent to Marine Biodiversity in the Caribbean Region

The following are the most important multilateral environmental agreements that guide global, regional and national action on environmental issues in the Caribbean region and have a significant impact upon the conservation and sustainable management of the Caribbean Region's marine biodiversity.

10.5.3.1 United Nations Convention on the Law of the Sea (UNCLOS)

UNCLOS serves as the principal legal foundation for the ecosystem management approach to the world's seas (Parris 2016; Singh 2008; UNCLOS 1982). As mentioned earlier, however, the UNCLOS ocean regime also hinders effective ecosystem management in the Caribbean region because it fragments jurisdiction over marine environments and resources through the implementation of maritime zones. Moreover, innocent, transit and archipelagic passage rights implemented within the Caribbean Sea result in substantial maritime traffic and ship-based pollution, which contribute to the deterioration and depletion of the region's marine environment.

10.5.3.2 Convention on Biological Diversity (CBD)

The Convention on Biological Diversity (CBD) is the principal international treaty related to the conservation and sustainable use of biodiversity (UNCBD 1992). The authorized governments of 27 of the 30 sovereign and non-sovereign islands within the CIBH region are parties to the CBD and the Nagoya Protocol (the United States is not a party to the CBD and its possessions include the non-sovereign territories of Puerto Rico and the US Virgin Islands, and the uninhabited island of Navassa) (CBD 2020a). The CBD is implemented through National Biodiversity Strategies and Action Plans (NBSAPs) that integrate biodiversity conservation and sustainable use into national legislation and policies, public education and participation, and the allocation of public investments, among other activities (CBD 2020b). During the 5th Conference of the Parties held in 2000, the CBD parties were called upon to adopt the ecosystem approach to their national legislation, policies and implementation activities (CBD 2000, 2004).

In 2010, the Strategic Plan for Biodiversity 2011–2020 (SPB) and the Aichi Targets were approved at the 10th Conference of the Parties (CBD 2010). Parties to the CBD were required to revise their NBSAPs to implement the SPB and the Aichi Targets. Of the 27 sovereign and non-sovereign islands within the CIBH that are party to the CBD, 23 have submitted revised NBSAPs following the adoption of the SPB, 3 sovereign islands submitted their first NBSAP prior to 2010 and are working on a revised version, and 1 sovereign island has not submitted an NBSAP but has advised the CBD Secretariat that activities are under way (CBD 2020c).

Parties to the CBD have reported on the status of their implementation of the SPB and the 20 Aichi Targets by submitting national reports (CBD 2020d). The sixth and most recent national reports were scheduled to be submitted by 31 December 2018 (CBD 2016). The authorized governments of 26 of the 30 sovereign and non-sovereign islands within the CIBH have submitted their sixth national report. Only the sovereign island of Grenada (party to the CBD) and the non-sovereign islands of Puerto Rico, the US Virgin Islands and uninhabited Navassa (US possessions not party to the CBD) have not done so.

Even though the authorized governments of CIBH countries have mainly complied with the preparation and submittal of national plans and reports under the CBD framework, the implementation of the SAP and the Aichi Targets will not be achieved in the Caribbean region by the scheduled implementation year of 2020. Among the factors that account for this result are: "lack of capacity and differing capacities among agencies and stakeholders; difficulties in encouraging sectors to take on responsibility for implementation; absence of interdisciplinary frameworks for integrating biodiversity into sectors; lack of adequate financing; and lack of interest and knowledge of biodiversity among key political and decision-making stakeholders" (CEPF 2019, p. 138). We will further expand on these obstacles below in Sect. 10.6.

10.5.3.3 Convention on Wetlands of International Importance Especially as Waterfowl Habitat (Ramsar Convention) and Convention on International Trade in Endangered Species of Wild Fauna and Flora (CITES)

The Ramsar Convention and CITES are two of the first international agreements to protect biodiversity ecosystems and species, including coastal ecosystems and marine species. The authorized governments of 26 of the 30 islands within the CIBH region are signatories to the Ramsar Convention (Ramsar Convention 1971), while only 1 CIBH country is not a party to CITES (CITES 1973).

10.5.3.4 International Convention for the Prevention of Pollution from Ships (MARPOL)

This treaty has been ratified by three sovereign island states within the CIBH region and by France, the Netherlands, the United Kingdom and the United States, which are sovereign over 18 island territories within the CIBH (IMO 2020a). Annex V to MARPOL (Prevention of Pollution by Garbage from Ships) defines ecologically rich marine areas with active sea traffic as "special areas" and provides them with a higher level of protection from ship-based pollution (MARPOL 1973). In 1991, the

"Wider Caribbean region including the Gulf of Mexico and the Caribbean Sea" was added to the list of special areas under Annex V (IMO 2020b).

10.5.3.5 Agreement for the Implementation of the Provisions of UNCLOS Relating to the Conservation and Management of Straddling Fish Stocks and Highly Migratory Fish Stocks (SFSHMFS)

This treaty has been ratified by five sovereign island states within the CIBH region and by France, the Netherlands, the United Kingdom and the United States, which are sovereign over 18 island territories within the CIBH (UN 1995). Furthermore, the fishing regime established by the treaty has influenced the governance of fisheries in the Caribbean region, including CARICOM's CCCFP among others (Singh 2008).

10.5.4 Regional Legal Frameworks: Multilateral Agreements Pertinent to Marine Biodiversity in the Caribbean Region

10.5.4.1 Cartagena Convention and Its Protocols

The most important regional agreement and legal framework for the implementation of an ecosystem management approach for the Caribbean Sea's shared marine resources is the Cartagena Convention and its Protocols (UNEP 2012a). The Cartagena Convention's framework includes the following technical protocols: the 1983 Oil Spills Protocol (UNEP 2012b), the 1990 Specially Protected Areas and Wildlife Protocol (UNEP 2012c) and the 1999 Pollution from Land-based Sources Protocol (UNEP 2012d). All three protocols have a direct impact on the conservation, protection and management of the WCR's marine biodiversity. As explained in Sect. 10.5.1.1, above, the Cartagena Convention is also the legal framework used by the UNEP RSP and the UNEP CAR/RCU. The authorized governments of 29 of the 30 sovereign and non-sovereign islands within the CIBH (excluding Haiti) have ratified the Cartagena Convention, thus committing to the joint protection, development and management of their common marine resources (UNEP 2020a). However, not all of the 29 have also ratified one or more of the three protocols mentioned above.

10.5.4.2 Agreement Establishing the Caribbean Community Common Fisheries Policy (CCCFP)

The CARICOM CCCFP treaty establishes the fishing regime framework for the conservation and sustainable management of fisheries and related ecosystems in the Caribbean region. Through the CCCFP, CARICOM countries address the challenge of illegal, unregulated and unreported fishing in the Caribbean Sea by providing significant protection to the region's marine environment. It is supported by CFRM's governing body, an intergovernmental fisheries organization (McConney et al. 2016b).

10.5.5 Non-Governmental Organizations (NGOs)

A quick survey by the Critical Ecosystems Partnership Fund has identified within the CIBH region "379 local, national, regional and international non-profit, non-governmental and academic organizations working on environment and conservation issues (including in the productive sectors of fisheries, agriculture and ecotourism)" (CEPF 2019). In many countries, these NGOs are responsible for much of the work being done in the region on biodiversity conservation and management (CARICOM 2018).

10.6 Challenges for Effective Marine Biodiversity Governance in the Caribbean Region

The 2011 SPB and Aichi Targets, scheduled for implementation by 2020, have not been achieved in the Caribbean region despite the overwhelming network of actors, policies and initiatives involved in the conservation and sustainable management of the Caribbean Sea's marine environment. Meanwhile, Caribbean marine biodiversity ecosystems, habitats and species are at a critical stage. Many reasons have been given to explain this dichotomy, some of which will be outlined below.

- Consensus exists as to the value of the LME approach for the development of marine conservation and sustainable management policies for the Caribbean Sea, and a long history of scientific research exists in the region. However, the current scientific knowledge about "Caribbean marine biodiversity and species distribution does not satisfy the needs for objectively defining geographic conservation priorities and designing management plans at a regional scale" (Miloslavich et al. 2010, p. 2). Moreover, critical scientific knowledge relevant to ocean governance in the WCR "has not been effectively communicated to decision-makers and the general public" (McConney et al. 2016a).

- The SIDS countries within the CIBH lack the capacity to implement and monitor the complex technical and scientific obligations and policies contained in the many international and regional environmental agreements to which they are parties to, and gaps exist in data and information (CEPF 2019; UNGA 2018; Fanning et al. 2009). "In cases where information exists, it is often not easily or readily accessible or available for region-wide decision-making" (Fanning et al. 2009, p. 226).
- The SIDS countries within the CIBH face a wide range of institutional challenges, such as "lack of awareness of the values of biodiversity, weak environmental legislation, shifting jurisdiction of environmental portfolios, overlaps and/or gaps in jurisdiction, limited technical information, inadequate human and technical capacity and insufficient funding for biodiversity" (CARICOM 2018, p. 1).
- The international, regional and national ocean governance structures in place overlap, compete among themselves and are uncoordinated, having been described as an "institutional maze" (Chakalall et al. 2007, p. 94; Pittman et al. 2015; Parris 2013). Also, there is a lack of a holistic integrated plan for the conservation and management of the Caribbean Sea, which results in fragmented and ineffective policies and initiatives (Agard and Cropper 2007). For instance, although hundreds of marine protected areas and marine reserves have been established in the Caribbean Sea since the 1980s, their impact on the region's marine biodiversity ecosystem has been limited due to the lack of coordinated management policies to control many of the local drivers outlined above (Kaplan et al. 2015; Guarderas et al. 2008). Furthermore, these international, regional and national ocean governance structures have diverse mandates, aspirations and perspectives, including, among others, on climate change, sustainable development, biodiversity, specially protected areas and wildlife, fisheries, wetlands, coral reefs, coastal zone management, land-based sources of pollution, ship-based sources of pollution, and tourism. Further complicating the situation is the fact that these governance structures themselves have overlapping membership (Fanning et al. 2009). The fragmented Caribbean ocean governance has been characterized as follows:

  > [A] diversity of networks of actors serving various purposes that seldom intersect effectively, but with the potential to do so if greater attention is paid to networking. Notably absent in most cases are interactions at the critical stage of communicating analysis and advice to shape coordinated decision-making.... [T]here is seldom a clear mandate by any national-, sub-regional-, or regional-level institution for management policies that address integration among sectors at levels up to the ecosystem scale of the CLME (Fanning et al. 2009, p. 224).

- Conservation and sustainable management of the Caribbean Sea is organized at the national or sub-regional levels instead of at a broader level more akin to an ecosystem approach (Agard and Cropper 2007).
- Natural disasters, collapsing economies, increasing poverty and inequalities, massive migration and, most recently, the Covid-19 pandemic have impacted the Caribbean region (particularly the CIBH countries), exacerbating national

governments' lack of capacity and ability to implement and monitor existing marine biodiversity policies.

10.7 Conclusion

As we look towards the future, accomplishing effective governance of the Caribbean Sea's biodiversity is as challenging as solving the myriad complex political, socio-economic and climate problems that confront the region. The following items encapsulate some minimum requirements to be included in an agenda for an effective marine biodiversity governance framework:

Essential scientific knowledge about marine biodiversity ecosystems and species in the Caribbean Sea needs to be further developed and effectively communicated to national and regional policy decision-makers in a timely and meaningful manner.

Technical and scientific capacities need further development at the national and local levels to increase the implementation, monitoring and enforcement of international, regional and national policies designed for the conservation, protection and sustainable management of Caribbean marine biodiversity resources.

Overall environmental awareness and sensitivities need to be increased throughout all sectors of society.

Regarding institutional arrangements and policies, they cannot continue to overlap and compete. What is required is a coordinated ecosystem approach to the conservation, protection and sustainable management of the Caribbean Sea. Local communities, NGOs and academia must be brought into the development, implementation and monitoring of marine biodiversity policies (see for example: Pittman et al. 2015; Parris 2013; Chakalall et al. 2007).

More importantly, biodiversity cannot be addressed in the Caribbean region without also promoting the sustainable development of essential infrastructure; economic activity that benefits and improves environmental conditions; resiliency; a reduction of poverty and inequalities; participatory democracy; the end of colonial rule in the region; the improvement of health and education systems; and natural disaster preparedness and response, among others.

Massive international investment will be required to repair the environmental, social, economic, political and democratic degradation in the region. It is through this restorative action that a healthy and vibrant Caribbean Sea can be envisaged. The future of the region depends on it.

References

ACS. Caribbean Sea Commission. (2020a). Retrieved September 28, 2020, from http://acs-aec.org/index.php?q=csc

ACS. Members and Associate Members. (2020b). Retrieved September 29, 2020, from http://acs-aec.org/index.php?q=content/members-and-associate-members

ACS. (n.d.). Convention Establishing the Association of Caribbean States. Retrieved September 28, 2020, from http://acs-aec.org/index.php?q=about/convention-establishing-the-association-of-caribbean-states

Agard, J. B. R., & Cropper, A. (2007). Caribbean Sea Ecosystem Assessment (CARSEA): A subglobal component of the millennium ecosystem assessment. *Journal of the Institute of Marine Affairs*, 1–85. Retrieved September 28, 2020, from https://www.cbd.int/doc/meetings/mar/rwebsa-wcar-01/other/rwebsa-wcar-01-crfm-03-en.pdf

Barreto-Orta, M., Méndez-Tejeda, R., Rodríguez, E., Cabrera, N., Díaz, E., & Pérez, K. (2019). State of the beaches in Puerto Rico after Hurricane Maria (2017). *Shore & Beach, 87*(1), 16–23. Retrieved September 28, 2020, from https://www.researchgate.net/profile/Rafael_Mendez-Tejada/publication/332182479_State_of_Beach_of_PR_2017/links/5ca5060292851c8e64b0e8c4/State-of-Beach-of-PR-2017.pdf

Branoff, B., Cuevas, E., & Hernández, E. (2018). Assessment of Urban Coastal Wetlands Vulnerability to Hurricanes in Puerto Rico. Retrieved September 28, 2020, from http://drna.pr.gov/wp-content/uploads/2018/09/FEMA-Wetlands-Report.pdf

CARICOM. (2001). Revised Treaty of Chaguaramas Establishing the Caribbean Community Including the CARICOM Single Market and Economy. Retrieved September 28, 2020, from https://caricom.org/documents/4906-revised_treaty-text.pdf

CARICOM. (2003). Agreement establishing the Caribbean Community Common Fisheries Policy (CCCFP). Retrieved September 28, 2020, from https://caricom.org/documents/12304-cccf-policy.pdf

CARICOM. (2018). The State of Biodiversity in the Caribbean Community: A Review of Progress Towards the Aichi Biodiversity Targets. ISBN 978-976-600-419-4. Retrieved September 30, 2020, from https://wedocs.unep.org/bitstream/handle/20.500.11822/26757/State_Biodiversity_ACP.pdf?sequence=1&isAllowed+y

CARICOM. (2020). Member States and Associate Members. Retrieved September 28, 2020, from https://caricom.org/member-states-and-associate-members/

Castro-Rivera, A., & López-Marrero, T. (2019). Allá viene el temporal: Major hurricanes in Puerto Rico between 1851 and 2019. *Caribbean Geography, 24*, 24–39. Retrieved September 28, 2020, from https://www.researchgate.net/publication/341554800_Alla_viene_el_temporal_Major_hurricanes_in_Puerto_Rico_between_1851_and_2019; https://www.tlopezmarrero.com/uploads/8/6/3/9/86396506/castro-rivera___lopez-marrero__2019__cg.pdf

CBD. (2000). Decision 5/6. Ecosystem Approach. Retrieved September 28, 2020, from https://www.cbd.int/decision/cop/?id=7148

CBD. (2004). Guidelines: The Ecosystem Approach. Retrieved September 28, 2020, from https://www.cbd.int/doc/publications/ea-text-en.pdf

CBD. (2010). Decision 10/2. The Strategic Plan for Biodiversity 2011–2020 and the Aichi Biodiversity Targets. Retrieved September 30, 2020, from https://www.cbd.int/decision/cop/?id=12268

CBD. (2016). Decision 13/27. National reporting. Retrieved September 30, 2020, from https://www.cbd.int/doc/decisions/cop-13/cop-13-dec-27-en.pdf

CBD. (2020a). List of Parties. Retrieved September 28, 2020, from https://www.cbd.int/information/parties.shtml

CBD. (2020b). National Biodiversity Strategies and Action Plans or Equivalent Instruments (NBSAPs). Retrieved September 28, 2020, from https://www.cbd.int/nbsap/

CBD. (2020c). Status of Development of National Biodiversity Strategies and Action Plans or Equivalent Instruments (NBSAPs) at 28 September 2020. Retrieved September 28, 2020, from https://www.cbd.int/nbsap/

CBD. (2020d). National Reports. Retrieved September 28, 2020, from https://www.cbd.int/reports/

CEPF. (2019). Ecosystem Profile: The Caribbean Islands Biodiversity Hotspot – Final Report. Retrieved September 28, 2020, from https://canari.org/wp-content/uploads/2017/08/cepf-caribbean-ecosystem-profile-december-2019.pdf

Chakalall, B., Mahon, R., McConney, P., Nurse, L., & Oderson, D. (2007). Governance of fisheries and other living marine resources in the Wider Caribbean. *Fisheries Research, 87*(1), 92–99. https://doi.org/10.1016/j.fishres.2007.06.009

CITES. (1973). Retrieved September 28, 2020, from https://www.cites.org/eng/disc/text.php

CRFM. (2010). Castries (Saint Lucia) Declaration on Illegal, Unreported and Unregulated Fishing. Retrieved September 29, 2020, from http://www.fao.org/fishery/docs/DOCUMENT/wecafc/15thsess/ref11e.pdf

Fanning, L., Mahon, R., & McConney, P. (2009). Focusing on living marine resource governance: The Caribbean large marine ecosystem and adjacent areas project. *Coastal Management, 37* (3–4), 219–234. https://doi.org/10.1080/08920750902851203

Fanning, L., Mahon, R., & McConney, P. (2013). Applying the large marine ecosystem (LME) governance framework in the Wider Caribbean Region. *Marine Policy, 42*, 99–110. https://doi.org/10.1016/j.marpol.2013.02.008

Guarderas, A. P., Hacker, S. D., & Lubchenco, J. (2008). Current status of marine protected areas in Latin America and the Caribbean. *Conservation Biology, 22*(6), 1630–1640. https://doi.org/10.1111/j.1523-1739.2008.01023.x

IMO. (2020a). List of Parties. Retrieved September 29, 2020, from https://treaties.un.org/Pages/showDetails.aspx?objid=0800000280291139

IMO. (2020b). Special Areas under MARPOL. Retrieved September 29, 2020, from http://www.imo.org/en/OurWork/Environment/SpecialAreasUnderMARPOL/Pages/Default.aspx

Kaplan, K. A., Ahmadia, G. N., Fox, H., Glew, L., Pomeranz, E. F., & Sullivan, P. (2015). Linking ecological condition to enforcement of marine protected area regulations in the greater Caribbean region. *Marine Policy, 62*, 186–195. https://doi.org/10.1016/j.marpol.2015.09.018

Lausche, B. (2008). Wider Caribbean Region—A pivotal time to strengthen regional instruments for biodiversity conservation. *The International Journal of Marine and Coastal Law, 23*(3), 499–530. https://doi.org/10.1163/092735208X331890

LMEH. (2020). A Regional Perspective on the World's Oceans. Retrieved September 28, 2020, from https://www.lmehub.net/

López-Marrero, T., & Castro-Rivera, A. (2019). Huracanes intensos en la cuenca del Atlántico y azotes en el Caribe insular: 1851 al 2019, *Centro Interdisciplinario de Estudios del Litoral*. 1–6. Retrieved September 28, 2020, from https://seagrantpr.org/communications-and-publications/other-publications/; https://www.tlopezmarrero.com/uploads/8/6/3/9/86396506/huracanes_intensos_en_la_ca_y_el_caribe_1851_al__2019.pdf

Mahon, R., Cooke, A., Fanning, L., & McConney, P. (2013). Governance arrangements for marine ecosystems of the Wider Caribbean Region. *CERMES Technical Report No. 60*. Retrieved September 28, 2020, from https://clmeplus.org/app/uploads/2019/10/CLME_Regional_Governance_Framework_report.pdf.pdf

MARPOL. (1973). Prevention of Pollution by Garbage from Ships (Annex V). Retrieved September 28, 2020, from https://treaties.un.org/doc/Publication/UNTS/Volume%201340/volume-1340-I-22484-English.pdf

McConney, P., Fanning, L., Mahon, R., & Simmons, B. (2016a). A first look at the science–policy interface for ocean governance in the Wider Caribbean Region. *Frontiers in Marine Science, 2* (119), 1–16. Retrieved September 30, 2020, from https://doi.org/10.3389/fmars.2015.00119

McConney, P., Phillips, T., Lay, M., & Nembhard, N. (2016b). Organizing for Good Fisheries Governance in the Caribbean Community (CARICOM). *Social and Economic Studies, 65*(1), 57–85. Retrieved September 30, 2020, from http://www.jstor.org/stable/26380110

Meyers, N., Mittermeier, R. A., Mittermeier, C. G., da Fonseca, G. A. B., & Kent, J. (2000). Biodiversity hotspots for conservation priorities. *Nature, 403*(6772), 853–858. https://doi.org/10.1038/35002501

Miloslavich, P., Díaz, J. M., Klein, E., Alvarado, J. J., Díaz, C., Gobin, J., et al. (2010). Marine biodiversity in the Caribbean: Regional estimates and distribution patterns. *PLoS ONE, 5*(8). https://doi.org/10.1371/journal.pone.0011916

Mittermeier, R. A., Robles Gil, P., Hoffmann, M., Pilgrim, J., Brooks, T., Mittermeier, C. G., et al. (2004). *Hotspots revisited: Earth's biologically richest and most endangered terrestrial ecoregions*. Mexico City, Mexico: CEMEX. ISBN: 968-6397-77-9.

Nieto-Blázquez, M., Antonelli, A., & Roncal, J. (2017). Historical biogeography of endemic seed plant genera in the Caribbean: Did GAARlandia play a role? *Ecology and Evolution, 7*, 1. https://doi.org/10.1002/ece3.3521

OECS. (2006). The St. George's Declaration on Principles for Environmental Sustainability in the Eastern Caribbean (Revised).. Retrieved September 26, 2020, from https://oecs.org/our-work/knowledge/library/social-development/st-george-s-declaration

OECS. (2010). Revised Treaty of Basseterre Establishing the Organisation of Eastern Caribbean States Economic Union. Retrieved September 26, 2020, from https://oecs.org/our-work/knowledge/library/legal/revised-treaty-of-basseterre

OECS. (2016). OECS Environmental Management Strategy. Retrieved September 28, 2020, from https://www.oecs.org/our-work/knowledge/library/social-development/revised-oecs

Parris, N. (2016). An ocean policy for the Wider Caribbean Region (WCR). *Social and Economic Studies, 65*(1), 7–56. Retrieved September 30, 2020, from https://www.jstor.org/stable/26380109

Parris, N. N. (2013). *Towards an Ocean Policy for Integrated Governance of the Caribbean Sea and the Sustainable Development of the Wider Caribbean Region (WCR): What could it look like and how would it work?*. Division for Ocean Affairs and the Law of the Sea. Office of Legal Affairs, United Nations, New York. Retrieved November 18, 2020, from https://www.un.org/Depts/los/nippon/unnff_programme_home/fellows_pages/fellows_papers/paris_1213_barbados.pdf

Pittman, J., Armitage, D., Alexander, S., Campbell, D., & Alleyne, M. (2015). Governance fit for climate change in a Caribbean coastal-marine context. *Marine Policy, 51*, 486–498. https://doi.org/10.1016/j.marpol.2014.08.009

Ramsar Convention. (1971). Retrieved September 28, 2020, from https://treaties.un.org/Pages/showDetails.aspx?objid=0800000280104c20

Robertson, D. R., & Cramer, K. L. (2014). Defining and dividing the greater Caribbean: Insights from the biogeography of shorefishes. *PLoS ONE, 9*, 7. https://doi.org/10.1371/journal.pone.0102918

Rodríguez-Rivera, L. E. (2014). Transportation of radioactive materials through the Caribbean Sea: the development of a nuclear-free zone. In D. D. Caron & H. N. Scheiber (Eds.), *The oceans in the nuclear age: Legacies and risks (Expanded Edition)* (pp. 169–195). Leiden, The Netherlands: Brill Nijhoff.

Scobie, M. (2012). Environmental justice and marine governance in the Caribbean. *IUCN Academy of Environmental Law eJournal, 1*, 30–41. Retrieved November 18, 2020, from http://www2.ecolex.org/server2neu.php/libcat/docs/LI/ANA-086442.pdf

Sherman, K. (1991). The large marine ecosystem concept: Research and management strategy for living marine resources. *Ecological Applications, 1*(4), 349–360. https://doi.org/10.2307/1941896

Singh, A. (2008). Governance in the Caribbean Sea: Implications for Sustainable Development. Retrieved September 28, 2020, from https://www.un.org/depts/los/nippon/unnff_programme_home/fellows_pages/fellows_papers/singh_0809_guyana.pdf

Spalding, M. D., Ravilious, C., & Green, E. P. (2001). *World Atlas of Coral Reefs* (pp. 147–177). Berkeley, CA: University of California Press.

UNDP and GEF. (2013). The Strategic Action Programme for the Sustainable Management of the Shared Living Marine Resources of the Caribbean and North Brazil Shelf Large Marine Ecosystems (CLME+ SAP). Retrieved September 26, 2020, from https://www.clmeproject.org/download/sap/?wpdmdl=3026&refresh=5f6a0c9d8fa6f1600785565

UNDP and GEF. (2020). The CLME+ Project. Retrieved September 26, 2020, from https://www.clmeproject.org/

UNEP. (1981). Intergovernmental Meeting on the Action Plan for the Caribbean Environment Programme. UNEP/CEPAL/IG.27/3 (16/Apr./1981). Retrieved September 25, 2020, from https://repositorio.cepal.org/bitstream/handle/11362/19462/S8100089_en.pdf?sequence=2&isAllowed=y

UNEP. (2012a). Convention for the Protection and Development of the Marine Environment of the Wider Caribbean Region (Carteagena Convention) and its Protocols. ISBN: 978-92-807-3255-9. Retrieved September 30, 2020, from https://wedocs.unep.org/bitstream/handle/20.500.11822/27875/SPAWSTAC5_2012-en.pdf?sequence=1&isAllowed=y

UNEP. (2012b). Protocol to the Convention for the Protection and Development of the Marine Environment of the Wider Caribbean Region Concerning Cooperation in Combating Oil Spills in the Wider Caribbean Region (24 Mar 1983). ISBN: 978-92-807-3255-9. Retrieved September 30, 2020, from https://wedocs.unep.org/bitstream/handle/20.500.11822/27875/SPAWSTAC5_2012-en.pdf?sequence=1&isAllowed=y

UNEP. (2012c). Protocol to the Convention for the Protection and Development of the Marine Environment of the Wider Caribbean Region Concerning Specially Protected Areas and Wildlife (18 Jan 1990). ISBN: 978-92-807-3255-9. Retrieved September 30, 2020, from https://wedocs.unep.org/bitstream/handle/20.500.11822/27875/SPAWSTAC5_2012-en.pdf?sequence=1&isAllowed=y

UNEP. (2012d). Protocol to the Convention for the Protection and Development of the Marine Environment of the Wider Caribbean Region Concerning Pollution from Land-based Sources and Activities (06 Oct 1999). ISBN: 978-92-807-3255-9. Retrieved September 30, 2020, from https://wedocs.unep.org/bitstream/handle/20.500.11822/27875/SPAWSTAC5_2012-en.pdf?sequence=1&isAllowed=y

UNEP. (2020a). Cartagena Convention. Retrieved September 26, 2020, from https://www.unenvironment.org/cep/who-we-are/cartagena-convention

UNEP. (2020b). Regional Seas Programme. Retrieved September 08, 2020, from https://www.unenvironment.org/explore-topics/oceans-seas/what-we-do/working-regional-seas/regional-seas-programmes

UNEP. (2020c). Regional Seas: A survival strategy for our oceans and coasts. Retrieved September 08, 2020, from https://unep.ch/conventions/info/seas/RSbooklet-E.pdf

UNEP. (2020d). Wider Caribbean. Retrieved September 26, 2020, from https://www.unenvironment.org/explore-topics/oceans-seas/what-we-do/working-regional-seas/regional-seas-programmes/wider

UNEP–WCMC. (2016). *The State of Biodiversity in Latin America and the Caribbean: A mid-term review of progress towards the Aichi Biodiversity Targets*. Cambridge, UK: UNEP–WCMC. Retrieved September 28, 2020, from https://www.cbd.int/gbo/gbo4/outlook-grulac-en.pdf

UNGA. (2016). Towards the sustainable development of the Caribbean Sea for present and future generations. A/RES/71/224. Retrieved September 26, 2020, from https://undocs.org/en/A/RES/71/224

UNGA. (2018). Sustainable Development of the Caribbean Sea for Present and Future Generations -- Report of the Secretary-General. A/73/225. Retrieved September 26, 2020, from https://undocs.org/en/A/73/225

United Nation (UN). (1995). Agreement for the Implementation of the Provisions of UNCLOS relating to the Conservation and Management of Straddling Fish Stocks and Highly Migratory Fish Stocks (SFSHMFS). Retrieved September 28, 2020, from https://treaties.un.org/Pages/ViewDetails.aspx?src=TREATY&mtdsg_no=XXI-7&chapter=21&clang=_en#EndDec

United Nations (UN). (1982). Convention on the Law of the Sea (UNCLOS). Retrieved September 30, 2020, from https://www.un.org/Depts/los/convention_agreements/texts/unclos/unclos_e.pdf

United Nations (UN). (1992). Convention on Biological Diversity (CBD). Retrieved September 25, 2020, from https://www.cbd.int/doc/legal/cbd-en.pdf

UNOHRLLS. (2020). List of SIDS. Retrieved September 25, 2020, from https://www.un.org/ohrlls/content/list-sids

US NOAA. (2018). Status of Puerto Rico's Coral Reefs in the Aftermath of Hurricanes Irma and Maria. Assessment Report Submitted by NOAA to the FEMA Natural and Cultural Resources Recovery Support Function. Retrieved September 26, 2020, from http://drna.pr.gov/wp-content/uploads/2018/06/Coral-Assessment-Report.pdf

US NOAA. (2020). Busy Atlantic hurricane season predicted for 2020: Multiple climate factors indicate above-normal activity is most likely. Retrieved September 26, 2020, from https://www.noaa.gov/media-release/busy-atlantic-hurricane-season-predicted-for-2020

Yu, M., Rivera-Ocasio, E., Heartsill-Scalley, T., Dávila-Casanova, D., Ríos-López, N., & Gao, Q. (2019). Landscape-level consequences of rising sea-level on coastal wetlands: Saltwater intrusion drives displacement and mortality in the twenty-first century. *Wetlands, 39*, 1343–1355. https://doi.org/10.1007/s13157-019-01138-x

Chapter 11
The Convention on Biological Diversity and the Mediterranean Sea: A Beacon for International Environmental Governance in the Region?

Guillaume Futhazar

Contents

11.1 Introduction ... 206
11.2 The Quasi Absence of the Mediterranean in the CBD Framework ... 209
11.3 The Limited Relevance of Systemic Integration for the Influence of the CBD on Mediterranean Regimes ... 210
11.3.1 A Brief Overview of Mediterranean International Environmental Law ... 210
11.3.2 The Limited Impact of Systemic Integration on Instruments of the Same Overall Regime ... 212
11.4 The CBD as an Emerging Guidepost for Biodiversity-Related Regimes in the Region 214
11.4.1 Norm Diffusion in International Law: A Key to the CBD's Influence ... 214
11.4.2 A Progressive Opening of Mediterranean Regimes to Norm Diffusion: A Path for the CBD's Influence ... 215
11.5 Conclusion ... 219
References ... 220

Abstract The Mediterranean region is a laboratory for international environmental law. As a biodiversity hotspot, it contains all of the familiar challenges of global environmental governance and it stands at the forefront of international legal cooperation, with initiatives such as the Mediterranean Action Plan and the General Fisheries Commission for the Mediterranean. Nowadays, the Mediterranean region's environmental agreements exist within a dense network of international institutions and conventions that shape their implementation and evolution. The purpose of this chapter will be to discuss the influence that the CBD has exerted, and still exerts, on the dynamics of international environmental law in the region.

The chapter will highlight that the formal legal influence of the CBD in the region is weak. The CBD does not directly address Mediterranean issues and it is unlikely to exert influence through the principle of systemic integration. However, it has

G. Futhazar (✉)
Max Planck Institut für ausländisches öffentliches Recht und Völkerrecht (Max Planck Institute for Comparative Public Law and International Law), Heidelberg, Germany
e-mail: futhazar@mpil.de

M. Campins Eritja, T. Fajardo del Castillo (eds.), *Biological Diversity and International Law*, https://doi.org/10.1007/978-3-030-72961-5_11

influenced international environmental law in the region through norm diffusion. This trend is likely to continue in the post-2020 framework, thus fostering greater coherence in international biodiversity law.

11.1 Introduction

> To look back at the history of the Mediterranean is to observe a symbiosis of man and nature that may be about to end (Abulafia 2011).

The current state of the Mediterranean environment is an unfortunate confirmation of David Abulafia's insight. Recent reports of the Mediterranean Action Plan (MAP) (MAP 2012, 2019a; Barcelona Convention 2017) clearly highlight that the region faces ever-increasing anthropic pressures. The Mediterranean is home to the busiest marine traffic routes in the world and has seen its coastlines drastically altered by rampant urbanization. In addition, overfishing still occurs in spite of international cooperation on the matter (FAO 2019).

These pressures put a unique environment at risk. Indeed, the Mediterranean region is a biodiversity "hotspot" (Cuttelod et al. 2008) with numerous endemic species, many of which are currently threatened with extinction. As a vulnerable, semi-enclosed sea, the Mediterranean Sea has been the object of early international cooperation in the field of environmental law (Haas 1990). With the coexistence of numerous international agreements in a region marked by an obvious North–South divide, the Mediterranean has indeed been a laboratory for international environmental law. For instance, the 1976 Barcelona Convention for the Protection of the Mediterranean Sea against Pollution was the first treaty to be adopted within the framework of the Regional Seas Programme of the United Nations Environmental Programme (UNEP), and it was one of the first framework conventions, which are a type of international agreement that is now a staple of international environmental law (Kiss 1993; Matz-Lück 2009). Similarly, the General Fisheries Commission for the Mediterranean (GFCM) is one of the earliest examples of regional cooperation in the matter of fisheries. Yet despite the existence of several international instruments pertaining to the Mediterranean environment and its natural resources, environmental degradation persists. In this regard, the Mediterranean is but another unfortunate example of the general ineffectiveness of environmental law.

This chronic ineffectiveness has been a powerful catalyst for legal innovations in the field. Having witnessed the failure of traditional legal methods (e.g. binding top-down measures limited to certain sectors) to generate the appropriate and needed change, the architects of international environmental law have turned to the design and implementation of new approaches (Maljean-Dubois 2017a; Beyerlin et al. 2006).[1] For instance, the 2015 Paris Agreement offers an example of the shift

[1] One could mention non-adversarial compliance mechanisms, which were first established in the context of multilateral environmental agreements.

from a binding top-down approach to a voluntary logic supported by a denser procedural framework (Klein et al. 2017).

Currently, the international regime[2] for the conservation and sustainable use of biodiversity is undergoing continuous change to meet the challenges of effectiveness. For biodiversity, 2020 has even been hailed as a "big year" (CMS 2020). Even though the pandemic has forced the big year to be rescheduled and the latest Global Biodiversity Outlook has issued grim conclusions on the current direction of biodiversity governance (CBD 2020a), the dynamics at play remain unchanged. In particular, the post-2020 framework for biodiversity (CBD 2020b) is close to completion and should be adopted during the next Conference of the Parties to the Convention on Biological Diversity (CBD). The new framework is expected to have a high impact on the governance of biodiversity as a whole, and it will likely serve as a reference for all actors involved. Given that the previous framework adopted within the CBD in 2010—the Strategic Plan for Biodiversity and its Aichi Targets (CBD 2010)—became a point of reference for most institutions and conventions dealing with biodiversity (Futhazar 2015), one could expect the same pattern to recur with the post-2020 framework.

Clearly, the CBD, while being only weakly normative,[3] has become a central element of international law on biodiversity. However, the modalities and extent of its influence vary depending on the regime in question. For instance, the influence of the CBD on the 1973 Convention on International Trade in Endangered Species of Wild Fauna and Flora (CITES) is based on elements (Cooney 2001) that differ from the CBD's potential impact on the United Nations Framework Convention on Climate Change (UNFCCC) (Maes et al. 2013). It is therefore interesting to unpack the role of the CBD in different legal contexts. This will be the purpose of the chapter in relation to the Mediterranean region, specifically its seas and coasts.

To understand the influence[4] of the CBD in the Mediterranean region, the chapter will be divided as follows. The next section examines the place of the Mediterranean region within the CBD and concludes that the CBD's direct influence on the region—understood here as the direct relevance of its provisions for the Mediterranean marine and coastal environment—is marginal at best. Indeed, the Mediterranean region is absent from the CBD's legal and institutional framework and is

[2]In this chapter, the word "regime" refers to the principles, rules, norms and procedures around which actors' expectations converge in a given field of international relations (Krasner 1983). This definition from international relations is used in order to broaden the scope of analysis. In this chapter, any reference to an agreement as a regime will imply that the different norms subsequently adopted for its implementation will be taken into consideration. Moreover, the actors in a regime will not only be the states, but also the numerous non-state actors, that contribute to maintaining the regime.

[3]Most of the articles of the convention start with the following wording: "Each Contracting Party shall, as far as possible and as appropriate . . . .".

[4]The notion of influence does not have a specific definition in law, political science or international relations. In the context of this chapter, the term is to be understood in its most general sense, *i.e.* "the power or capacity of causing an effect in indirect or intangible ways" (Merriam – Webster).

mentioned only in the context of a technical process that aims to identify important ecological areas.

Given the CBD's weak direct influence, the following sections try to unpack its indirect influence. In the context of the chapter, indirect influence is understood as any influence that the CBD may have on other international rules applicable to the Mediterranean environment. This influence may be on the interpretation and implementation of existing norms, or on the design of new ones. These are mainly the rules contained in the Barcelona Convention and its related protocols, the General Fisheries Commission for the Mediterranean, and the Agreement on the Conservation of Cetaceans of the Black Sea, Mediterranean Sea and contiguous Atlantic Area (ACCOBAMS).

The third section then briefly lays out these regimes and examines the CBD's indirect influence through a law of the treaties perspective. The CBD and the different regional instruments applicable to the Mediterranean environment coexist in the same legal order, and as such are related to one another. This implies a potential for systemic integration (Pasquet 2018; McLachlan 2005), which could lead to considerable weight being given to the CBD in the interpretation of international instruments in the Mediterranean region. However, as the section will also show, systemic integration is unlikely to generate noticeable effects as the Mediterranean agreements in force in the region are *lex specialis* in comparison to the CBD. They already contribute to the fulfillment of some of its purposes.

Next, the fourth section examines a form of influence which, in contrast to the ones described in the previous sections, has the potential to generate tangible effects: norm diffusion (Maljean-Dubois 2017b; Delcour and Tulmets 2019; Ruiz Fabri and Gradoni 2009). The section highlights how norms and concepts generated within the CBD framework have subsequently been endorsed by the institutions and parties of the Mediterranean agreements, thus leading to indirect influence from the CBD in the region.

Lastly, the conclusion discusses the ever-growing centrality of the CBD for all biodiversity-related conventions, both globally and regionally. It has become a beacon for this branch of international law. As such, therefore, the shift may be seen as an additional sign of the constitutionalization of international environmental law.

In sum, the chapter argues that the CBD influences international environmental law in the region through informal legal ways. Moreover, its influence is likely to continue as we enter into the post-2020 framework. While this is a positive development for the coherence of international environmental law, it unfortunately fails to address the structural problems that have impeded the CBD's effectiveness in the region.

It should be stressed, however, that this conclusion is dependent on the methodology used for this chapter. Indeed, the influence of the CBD is explored here through the analysis of official documentation made available by public actors and through reference to the literature on the topic. This approach is potentially blind to other forms of influence that might be witnessed if one chose to focus, for instance, on actors rather than on laws and institutions. However, if different conclusions were reached through a different methodology, this discrepancy would in itself highlight the different path of the influence of the CBD. Therefore, more research, and particularly interdisciplinary research, should be conducted to fully unpack the influence of the CBD in the Mediterranean.

11.2 The Quasi Absence of the Mediterranean in the CBD Framework

Like other biogeographic regions, the Mediterranean region is absent from both the text of the CBD and its institutional structure. While a brief mention of the Mediterranean can be found in recent decisions of the Conference of the Parties to Ecological or Biologically Significant Marine Areas (EBSA), these decisions have no normative content and are intended only to provide information and technical assistance.

The EBSA process was initiated to address biodiversity in areas beyond national jurisdiction when the issue of biodiversity became more pressing on environmental and maritime agendas. In 2006, the parties to the CBD met in Curitiba and adopted a decision concerning protected areas (CBD 2006). In the decision, the parties insisted on the central role of the UN General Assembly "in addressing issues relating to the conservation and sustainable use of biodiversity in marine areas beyond national jurisdiction". They also highlighted that the United Convention on the Law of the Sea (UNCLOS) is the primary legal framework in which such an issue should be tackled. In other words, the decision managed the overlap between the biodiversity regime and the law of the sea regime by attributing legal primacy to a specific instrument.

Nevertheless, the parties to the CBD stressed that the topic had to be approached in a spirit of inter-regime cooperation. In this respect, the parties decided that the CBD should be a tool for the provision of technical and scientific information relating to marine protected areas in areas beyond national jurisdiction. The result of this approach was the identification of Ecological or Biologically Significant Marine Areas or EBSAs.

EBSAs are "geographically or oceanographically discrete areas that provide important services to one or more species/populations of an ecosystem or to the ecosystem as a whole, compared to other surrounding areas or areas of similar ecological characteristics" (CBD 2008). In other words, the areas in question are key zones that should be protected because of their particularities. In 2008, the parties to the CBD adopted a series of criteria to allow for their identification: uniqueness or rarity; special importance for life history stages of species; importance for threatened, endangered or declining species and/or habitats; vulnerability, fragility, sensitivity or slow recovery; biological productivity; biological diversity; and naturalness (CBD 2008). In 2014, the parties to the CBD recognized the first 15 Mediterranean EBSAs. Their identification was made following a workshop in

Malaga, Spain, which involved participants affiliated to the Mediterranean Action Plan (CBD 2014).

The EBSA process is an illustration of the explicit management of regime interaction (Young 2012). While the legal approach was left within the realm of UNCLOS, the CBD took on the scientific and technical aspects of protection in areas beyond national jurisdiction. However, this organization implies that the designation of Mediterranean EBSAs carries no direct legal consequences. From a normative standpoint, the EBSA process is intended to be strictly descriptive, with some elements of technical assistance, whereas it leaves any decision to actually adopt regulations to ensure the areas' conservation to other forums. Admittedly, if the areas are "special", it logically ensues that they should benefit from the protection of an international legal system in which the principle of sustainable development is gaining traction (Barral 2012). So there is arguably at least some residual normativity in identifying EBSAs. Beyond that, however, the legal relevance of EBSAs in terms of conservation and sustainable use of the Mediterranean marine environment is minor.

Though the Mediterranean is absent from the CBD framework, the convention can exert its influence by other means. Notably, the CBD can play a role in the interpretation of Mediterranean obligations. Indeed, we are now witnessing a diffusion of norms from the CBD to the Mediterranean regimes. The following sections explore these aspects.

11.3 The Limited Relevance of Systemic Integration for the Influence of the CBD on Mediterranean Regimes

Before turning to the limits of systemic integration, it is important first to lay out the various international agreements and organizations that are relevant to the Mediterranean region's marine and coastal environment.

11.3.1 A Brief Overview of Mediterranean International Environmental Law

When considering the sustainable use and conservation of the Mediterranean environment, the first and most obvious international agreement is the Barcelona Convention. The convention was adopted in 1976 as the first achievement of the Mediterranean Action Plan, which had been instigated a year earlier (Haas 1990). Initially, it was called the "Barcelona Convention for the Protection of the Mediterranean Sea against Pollution". As the name suggests, the purpose was to protect the Mediterranean Sea from pollution (both from ships and land-based). To this end, the convention focused mainly on the marine environment. Over the years, however, its scope has expanded through the addition of protocols on a range of topics, such as specially protected areas, pollution from ships and land-based pollution.[5] In 1994, the convention was amended to cover the coastal environment. It was then renamed the "Convention for the Protection of the Marine Environment and the Coastal Region of the Mediterranean" so as to better reflect its broader scope and purpose. As with most modern multilateral environmental agreements, the parties to the Barcelona Convention meet on a regular basis to discuss the implementation of the convention and its protocols and adopt decisions on their strategic orientation. The decisions adopted at the meetings are elaborated by the MAP's permanent institutions, which gather the necessary expertise to fulfil this aim. By following this process, the parties and institutional actors allow for the adaptation and evolution of their convention framework.

The General Fisheries Commission for the Mediterranean is a central organization for the management of fisheries in the region. Established in 1949 under the auspices of the Food and Agriculture Organization (FAO), the fisheries organization marked the region's first attempt at international cooperation in the management of a shared natural resource. At the time, it was named the General Fisheries Council for the Mediterranean and had merely consultative powers. It exerted only marginal influence on the exploitation of fisheries in the region until a reform in 1997, when it was endowed with an autonomous budget and the power to issue binding decisions and it was renamed as a commission (Peñas Lado 2016; Tavares de Pinho 1997). The members of the GFCM meet on a yearly basis to take decisions on fisheries. These decisions are prepared by the technical bodies of the GFCM, the scientific advisory committee and the committee on aquaculture.

Last comes ACCOBAMS. This agreement was adopted in 1996 to implement the provisions of the Bonn Convention on Migratory Species in the region, specifically for cetaceans (Burns 1998). The agreement's main purpose is to prevent the deliberate catch of cetaceans in the region and establish strategic plans to keep the cetacean population at a sustainable level of conservation. Here again, the parties meet on a regular basis, every two years, to adopt decisions regarding the implementation of the agreement. It should be noted that ACCOBAMS's scope is broader than the Mediterranean Sea alone. It also covers the contiguous Atlantic area and the Black Sea, and in fact some of its members are not Mediterranean States (for example, Portugal and Georgia).

The three instruments described above overlap greatly and, as such, have come to co-evolve in a coordinated manner in recent years (GFCM 2016; Barcelona Convention 2016c; ACCOBAMS 2016a). However, they also exist in complementarity with the CBD, and allow for the accomplishment of its objectives at the regional level. Moreover, all the parties to the Barcelona Convention, GFCM and ACCOBAMS are parties to the CBD. This legal overlap raises questions of systemic

---

[5]The complete list of protocols is available on the MAP website at https://www.unenvironment.org/unepmap/who-we-are/barcelona-convention-and-protocols (last accessed on 1 October 2020).

integration, notably through the implementation of article 31.3.c of the Vienna Convention on the Law of Treaties (VCLT).

11.3.2 The Limited Impact of Systemic Integration on Instruments of the Same Overall Regime

Systemic integration simply refers to the notion that "international law norms should always be read as parts of a broader legal system" (Pasquet 2018; McLachlan 2005). This seemingly straightforward idea was not always a given. For instance, the International Court of Justice (ICJ) famously referred to some branches of international law as "self-contained regimes" (ICJ 1980). In its report on the fragmentation of international law, however, the International Law Commission (ILC) dismissed the idea of a truly self-contained regime and stressed that international law had a systemic nature (ILC 2006) and that its rules were meant to be read against a background of general international law and other relevant rules.

Article 31.3.c of the VCLT embodies this notion and indicates that, when interpreting a treaty, "any relevant rules of international law applicable in the relations between the parties" shall be taken into account. This paragraph of article 31 VCLT on interpretation was described by then Judge Xue Hanqin as a "master key" for the defragmentation of international law (ILC 2006). In the hands of judges, it has produced significant legal effects, such as taking into account international custom to delimit the potential meaning of provisions in a bilateral treaty (ICJ 2003). However, examining the potential use of article 31.3.c in the Mediterranean context poses two inherent limits to this interpretative approach.

The first and most obvious limit is that this provision of the VCLT will generate visible effects only within the context of a legal dispute. In the Mediterranean, the various international agreements specific to the region's environment have rarely been invoked in the context of legal disputes, the only well-known example being the Étang de Berre case brought before the European Court of Justice (ECJ) (ECJ 2004).

Even if disputes were common among parties to the Mediterranean region's environmental agreements, however, the principle of systemic integration would be unlikely to increase the CBD's influence on the region's agreements. Indeed, the principle of systemic integration seems to produce noticeable effects only when rules with clearly different designs and rationales are set against one another. Examples would include a specific bilateral rule and custom (ICJ 2003) or a rule of international trade with a principle of international environmental law (WTO 1998). In such cases, the function of systemic integration as a tool to ensure the systemic nature of international law becomes apparent. However, within a branch of international law with a clear purpose, such as biodiversity-related conventions, the added value of article 31.3.c for the coherence of law is less obvious.

This is due to two elements. First, article 31.3.c calls for an articulation of formal rules of international law (Bhat 2019; ILC 2006), thus leaving aside informal norms. Second, a rule with a similar design and purpose to the one being interpreted will not contribute additional meaning. The following Mediterranean examples show how these elements are limiting.

The Barcelona Convention and its protocols have numerous articles that contain words such as "necessary" and "appropriate". Such semantic choices allow parties to agree on the wording during negotiations (Perelman and Vander Elst 1984; Kulick 2016) and give the agreement an evolutive nature (Fitzmaurice 2008, 2009). What is necessary for the conservation and sustainable use of the environment will vary depending on the state of the environment and available knowledge. For instance, article 13 of the protocol on biological diversity indicates that "[t]he Parties shall take all appropriate measures to regulate the intentional or accidental introduction of non-indigenous or genetically modified species to the wild [...]". Should a dispute arise in the matter of the introduction of non-indigenous species in the region, this article, and particularly the notion of appropriateness, could be interpreted by referring to the CBD. Indeed, article 8.h of the CBD provides that "each Contracting Party shall, as far as possible and as appropriate [...] prevent the introduction of, control or eradicate those alien species which threaten ecosystems, habitats or species". It is therefore a relevant instrument on the question of non-indigenous species. However, this article of the CBD fails to add meaning to article 13 of the protocol. It simply reiterates a similar idea of control of invasive alien species (Burgiel 2017; Riley 2010). Here, it appears that article 31.3.c VCLT was not intended to bring coherence to similar provisions. To determine what exactly is meant by the word "appropriate", other documents are required, such as technical reports, soft norms adopted during conferences of the parties, and standards (CBD 2018), none of which fall within the categories referred to by article 31.3.c VCLT. They could perhaps be seen as "context", in accordance with article 31.1 VCLT.[6] However, the documents produced within the CBD framework are not necessarily the most relevant, given that the members of the MAP have adopted guidelines specifically for the Mediterranean region (MAP 2008).

The same pattern recurs in the context of ACCOBAMS and the General Fisheries Council for the Mediterranean. The appropriate measures to conserve the region's cetacean populations will not be made clearer by referring to any provisions on *in situ* conservation set out in the CBD. Nor will the CBD help to determine what amounts to the long-term sustainability of fishing activities in the context of the GFCM. Because the rationale of the CBD—at least in terms of conservation and sustainable use—already permeates these regimes, systemic integration provides no obvious added value. When it comes to interpreting the Mediterranean agreements, the CBD is simply not the most relevant point of reference.

[6]"A treaty shall be interpreted in good faith in accordance with the ordinary meaning to be given to the terms of the treaty in their context and in the light of its object and purpose" (emphasis added).

At this stage in the chapter, the importance of the CBD for Mediterranean seems to be rather marginal. When it comes to the Mediterranean environment, the existing regional agreements constitute a *lex specialis* that does not warrant the use of the CBD *per se* to conserve and sustainably use the region's environment. Here, however, the notion of *lex specialis* does not imply a conflict between general and special provisions, but rather a state of complementarity (Banaszewska 2015). The Mediterranean agreements complement the provisions of the CBD for the region and therefore push the CBD into the background, only to fill normative gaps if needed.

However, this place in the "background" does not totally preclude any legal effects of the global convention in the region. To state as much would be to ignore that the CBD also exerts its leverage through soft strategic influence. This has been the case among the other multilateral environmental agreements of the biodiversity-related convention in the Mediterranean region.

This influence may be indicative of a continuing constitutionalization of international law on biodiversity led by the CBD as a weak normative instrument with considerable pull in terms of concepts and telos.

11.4 The CBD as an Emerging Guidepost for Biodiversity-Related Regimes in the Region

This section argues that the CBD influences the development of Mediterranean agreements through international norm diffusion (Maljean-Dubois 2017b; Delcour and Tulmets 2019; Ruiz Fabri and Gradoni 2009, Ovodenko and Keohane 2012). To this end, the section begins with a brief introduction to the notion of norm diffusion before turning to its concrete manifestations within Mediterranean regimes.

11.4.1 Norm Diffusion in International Law: A Key to the CBD's Influence

Norm diffusion is a broad notion in comparative law that covers many different labels (transplants, reception, spread, circulation...) used to describe how norms can circulate from one order to another (Twining 2013; Parks and Morgera 2015). Generally, diffusion will refer to occurrences "when one legal order, system or tradition influences another in some significant way" (Twining 2013). Though comparative law has traditionally focused on States as its main unit of analysis, its scope has gradually broadened over the year to include different types of legal orders (Morgera 2015). Recent studies have even focused on norm diffusion across international regimes (Maljean-Dubois 2017b) and have highlighted that norm diffusion across regimes amounts to a form of regime interaction (Young 2012) that can foster coherence under certain conditions (for instance, if the diffusion was concerted) (Maljean-Dubois 2017b).

With regards to norm diffusion across regimes, several types of diffusion can be observed.

For instance, a norm adopted in a regime can be adopted in another one with the same form and substance. This is the case when fishing regulations adopted by the European Union are also adopted by the members of the GFCM. Such occurrences amount to normative mimicry, and are a fairly common phenomenon. In national contexts, these are sometimes referred to as "legal transplants" (Miller 2003; Fedtke 2006). The mimicry can be acknowledged, with explicit reference to the source of the norm, or it can be tacit, with similar norms occurring in different contexts, but without cross-references.

However, other forms of diffusion which do not amount to a transplant can also occur. Members of a regime can specify that the norms they have adopted will contribute to the fulfilment of the purposes of other regimes. This approach streamlines the effort of different regimes in order for them to contribute simultaneously to a common purpose. This was the case from 2010 to 2020 when several biodiversity-related conventions adopted the Aichi Targets as their own (Futhazar 2015; Rogalla von Bieberstein et al. 2018), and it is the case when EU member states include references to the European Convention on Human Rights in their charter of fundamental rights (Amalfitano 2018). In these cases, it was not norms that circulated *per se*, but their purposes (Buffard Zemanek 1998; Futhazar 2015).

11.4.2 A Progressive Opening of Mediterranean Regimes to Norm Diffusion: A Path for the CBD's Influence

The Mediterranean regimes studied in this chapter do not exist in isolation. Their members and institutions are attentive to the evolution of environmental law in other regions and at the global level.

Their history reflects this tendency in telling ways. For instance, following the 1992 Rio Conference, the parties to the Barcelona Convention amended the treaty to incorporate the "new" environmental principles recognized by the international community at the beginning of the decade. The amended Barcelona Convention now makes reference to the prevention principle, the "polluter pays" principle and the precautionary approach/principle (Scovazzi 1999). The protocol concerning Specially Protected Areas and Biological Diversity, even makes reference in its preamble to the CBD as an obvious point of reference for its implementation.

Similarly, ACCOBAMS, as an instrument of the CMS family, has obvious ties to the Bonn Convention and is aligned with its purposes and strategic objectives.[7] Also, the GFCM as a regional fisheries management organization embedded within the

[7]See article V CMS - Guidelines for AGREEMENTS.

institutional framework of the FAO is prompt to incorporate the new policies of the FAO (GFCM 2011b).

More pertinently, the CBD's influence can be seen in recent decisions of the Mediterranean regimes. These include COP decisions, resolutions and recommendations that have been adopted by the Mediterranean states, but prepared by a multitude of actors: members of permanent institutions, appointed experts and NGOs that constitute a network that captures the essence of the regime they constitute. As such, the decisions are an expression of the regime as a whole, and of its architects. They are also instrumental in interpreting the obligations contained in the different Mediterranean agreements. Seeing the place held by the CBD in these documents is as such indicative of the dynamic that exists between the regimes beyond the coexistence of black letter international law.

Studies of recent Mediterranean decisions show that there is a progressive incorporation of the CBD's objectives and that the post-2020 framework is intended to be a reference for the region's regimes in the years ahead. This trend will thus reinforce the CBD's place as the central pillar of international biodiversity law in the coming years. However, the attitudes of the Mediterranean regimes towards the CBD have been heterogeneous. In particular, the GFCM has remained primarily affiliated to other fisheries regime.

11.4.2.1 An Initial Self-Referential Attitude Within the Barcelona Convention

While amendments have been made to the Barcelona Convention to modernize the treaty by incorporating new legal tools, the attitude of the regime's architects appears to have been mainly self-referential until fairly recently. This is visible for instance in the initial strategic action plan for biodiversity (SAP BIO) adopted in 2003 (MAP 2003). The action plan was intended to be a core framework for the coherent implementation of the Biodiversity Protocol (MAP 2019b). It is complemented by more specific action plans for certain species and ecosystems (Barcelona Convention 2008, 2019). Even though these documents amount to soft law, they are nevertheless key to interpreting the potential meaning of the vague obligations contained in the Barcelona Convention and its protocols. Any references that they make to other agreements can therefore shape the meaning of the Mediterranean region's obligations.

However, an analysis of the SAP BIO and its first action plans shows that they contain only passing references to other relevant multilateral environmental agreements. The SAP BIO stresses the importance of synergies with other agreements but does not go beyond that. The only notable exception is an invitation to implement the Ramsar Convention on wetlands insofar as to identify important wetlands in the region. The specific plans, such as the plans for monk seals and turtles, operate in a similar fashion. For instance, the plan on turtles could have made a useful reference to the CITES annexes that also cover the species present in the region (Barcelona

Convention 2008), but it makes no such mention. The overall logic of these documents was that they were self-standing.

This approach has changed in recent decisions on the implementation of the Barcelona Convention. Specifically, these decisions have shed light on the existing networks of international rules that are relevant for the Mediterranean environment. They are more connected and open than the action plans mentioned above. For instance, in 2016, the parties to the Barcelona Convention adopted a decision for the specific purpose of meeting Aichi Target 11[8] for marine protected areas in the region (Barcelona Convention 2016a). Also, in their most recent plan on invasive alien species (Barcelona Convention 2016b), there are extensive references to other norms and standards for the management of the environmental threat (Burgiel 2017; Riley 2010).

The recent negotiations on the future of the SAP BIO leave no doubt about the influence of the CBD in the coming years. The latest evaluation of the SAP BIO clearly indicates that "it is crucial to ensure maximum harmonization between the new orientations and priorities to be promoted in the post-2020 SAP BIO and those that will be decided at the global level in the post-2020 Biodiversity Framework to be adopted in October 2020 by the CBD" (MAP 2019b). To this end, the parties to the Barcelona Convention have explicitly requested the Secretariat of the Convention "to prepare in 2020-2021 the 'Post-2020 Strategic Action Programme for the Conservation of Biodiversity and Sustainable Management of Natural Resources in the Mediterranean Region' (Post-2020 SAP BIO), aligned with the Sustainable Development Goals, harmonised with the CBD Post-2020 Global Biodiversity Framework through the optic of the Mediterranean context" (Barcelona Convention 2019).

11.4.2.2 An Open Approach Within ACCOBAMS

The attitude of the Barcelona Convention regime stands in contrast to ACCOBAMS, which frequently makes reference to other global regimes that are relevant for its purposes.

For instance, the parties to ACCOBAMS often refer to other multilateral environmental agreements in their resolutions. Several ACCOBAMS resolutions mention CITES as an important contextual element for their implementation and even invite the ACCOBAMS parties to take into account COP Decisions of CITES (ACCOBAMS 2002a, b, 2004, 2007a, 2013a). ACCOBAMS resolutions also take into account the technical work done by other relevant institutions. The 2016 resolution on species conservation and management plans (ACCOBAMS 2016b)

[8] By 2020, at least 17 per cent of terrestrial and inland water areas and 10 per cent of coastal and marine areas, especially areas of particular importance for biodiversity and ecosystem services, are conserved through effectively and equitably managed, ecologically representative and well-connected systems of protected areas and other effective area-based conservation measures, and integrated into the wider landscape and seascape.

explicitly acknowledges the guidelines prepared by the International Whaling Commission.

In this spirit of connectedness, the CBD often appears in ACCOBAMS resolutions. As early as 2007, the parties to ACCOBAMS adopted a decision on marine protected areas indicating that it was intended to contribute to the strategic plan of the CBD (ACCOBAMS 2007b). This trend continued after the adoption of the Strategic Plan for Biodiversity and its Aichi Targets (ACCOBAMS 2013b). Even before the adoption of the post-2020 framework, the parties to ACCOBAMS adopted a resolution that made clear that the upcoming framework would guide the implementation of the agreement (ACCOBAMS 2019). Much as in the context of the Barcelona Convention, the Mediterranean states are anticipating the new global framework and adjusting their regimes accordingly.

11.4.2.3 A Focused Approach Within the GFCM

In comparison to ACCOBAMS, the GFCM appears to have a more focused approach when it comes to incorporating elements from other regimes. Specifically, the GFCM remains largely focused on the conceptual and legal innovations that emanate from the FAO. For instance, the code of conduct for responsible fisheries (FAO 1995; Friedrich 2008) has been an obvious influence on many of the decisions taken by the members of the commission in recent years (GFCM 2015). The CBD, though occasionally mentioned in the working documents of the GFCM (GFCM 2011a), does not seem to have had the same influence as it has had on the Barcelona Convention or ACCOBAMS. This lack of influence indicates that the GFCM, though part of the regime for the conservation and sustainable use of the Mediterranean, is not necessarily part of the same broader legal family as ACCOBAMS and the Barcelona Convention. Both of the latter two agreements lie squarely within the realm of international environmental law, with a clear focus on conservation. By contrast, the GFCM, even though it has incorporated sustainable development as an important concept in its framework, remains above all an organization dealing with fisheries, i.e. a harvesting activity for trade and consumption purposes.

In sum, there is an obvious link between the pattern of norm diffusion that we can see in the region, and the telos of the regime incorporating new norms. Consequently, the Barcelona Convention and ACCOBAMS appear to be acting as a sort of regional relay station for the CBD by adapting its goals to the region's specificities and species. While the GFCM shows no signs of norm diffusion from the CBD, it nevertheless works in close collaboration with the Barcelona Convention and ACCOBAMS. The CBD's strong influence on the latter two regimes may bring some influence to bear on legal development within the GFCM.

11.5 Conclusion

This overview of the recent evolution of Mediterranean regimes leaves us with the following assessment: the CBD's role as a compass is growing, at least within the contexts of the Barcelona Convention and ACCOBAMS. In contrast, the GFCM, which is less of an environmental regime and more of a resource management regime, seems to remain outside of the CBD's growing influence.

In other words, the Barcelona Convention and ACCOBAMS are contributing to some of the CBD's goals and their respective parties have expressed their intention to continue in this direction as we embark on the post-2020 framework. This regional attitude is similar to what is already happening at the global level within other multilateral environmental agreements. The goals of the CBD are becoming truly global goals by being disseminated in other forums.

This circulation of goals evokes an initial draft of the CBD before its adoption in 1992. A version of the text, distributed in 1991 (UNEP 1991), envisioned that existing environmental agreements could be renegotiated in order to become protocols to the CBD.[9] Convention–protocol pairing is a frequent legal technique in international environmental law. A protocol will typically allow for the implementation of a convention on a specific topic (Kiss 1993; Matz-Lück 2009). However, there is no special legal relationship between a convention and a protocol beyond their explicit affiliation. For instance, becoming a party to a protocol is not necessarily conditioned by prior ratification of the convention to which it is affiliated.[10] A framework convention will serve as an obvious reference point for the interpretation of an affiliated protocol, but there are no further legal consequences.

If this is the case, the increasingly widespread appeals to the CBD and its objectives in the decisions and frameworks of other multilateral environmental agreements could be seen as a way to implement former article 21.2 without ever mentioning it. In other words, existing conventions dealing with biodiversity, both globally and regionally, are becoming tools to implement the CBD's purposes on specific issues.

This centrality of the CBD could be described as a weak form of constitutionalization in international environmental law. The constitutionalization of international

[9]Article 21.1: "any existing international agreement relating to the conservation and sustainable use of biological diversity may be renegotiated as protocols to the present Convention".

[10]Some framework conventions can specify this condition, but in the absence of such precision in the framework convention or its protocol, there is no general principle limiting access to a protocol only to states that are parties to the framework convention (despite some commentators arguing otherwise—Matz-Lück 2009). For instance, in the Mediterranean region, Turkey is not a party to the CMS and has nevertheless ratified ACCOBAMS. In contrast, article 29.1 of the Barcelona Convention specifies that

> No one may become a Contracting Party to this Convention unless it becomes at the same time a Contracting Party to at least one of the protocols. No one may become a Contracting Party to a protocol unless it is, or becomes at the same time, a Contracting Party to this Convention.

law refers to the fact that international law is moving from a system based on traditional principles of sovereignty, consensualism and territorial integrity to a system based on incorporated norms inspired by constitutionalism (Peters 2017). In this shift, some fields appear to have witnessed the emergence of norms that serve a constitutional function: "to establish, organize, integrate and stabilize a political community, to contain political power, to provide normative guidance and to regulate the governance activities of lawmaking, law application, and law enforcement." (Peters 2017). At this time, however, we have not witnessed the emergence of a supra-constitution that fully fulfils the function described above in the international sphere (Klabbers et al. 2011). Rather, we observe "constitutional fragments", that is, signs of constitutionalization in different sub-fields of international law and even beyond the traditional public structures of law (Teubner 2012). In international environmental law, however, the phenomenon is still weak. It has been well established that the field lacks a constitution (Bodansky 2009) and that there is currently no serious attempt to establish one (Kotzé 2019). Instead, international environmental law possesses a set of soft norms that entirely permeate it, with some of these soft norms solidifying into harder formal legal rules (Krämer and Orlando 2018).[11] The diffusion of the CBD's goals potentially adds another layer to the process as it harmonizes the purpose of existing norms. In this sense, the diffusion of norms emanating from the CBD can help to stabilize a political community and provide normative guidance. Of course, this phenomenon lacks the ability to contain and regulate political power, and is thus far from bringing about a global environmental constitution on its own. It is nevertheless an expression of a shared desire within a political community to work towards a set on commonly agreed goals.

As the environmental crisis worsens with each passing year, such an attempt to create greater international legal coherence is welcome. However, one may wonder whether the lack of coherence in international biodiversity law is what has actually been hindering its implementation.[12] In order for the next decade to end on a positive environmental assessment, both globally and in the Mediterranean region, additional reforms will be needed. These reforms will potentially be much harder to accomplish than simply bringing coherence to a pre-existing legal system.

References

Abulafia, D. (2011). *The Great Sea: A human history of the Mediterranean*. London: Penguin Books.

ACCOBAMS. (2002a). Resolution 1.10, Cooperation between National Networks of Cetacean Strandings and the Creation of a Database.

[11]This was the case with the rules on impact assessments (ICJ 2010).

[12]For instance, the subsidies for environmentally harmful activities such a fossil fuels extraction and use have been rising in the past years despite calls for their reduction (IPBES 2019).

ACCOBAMS. (2002b). Resolution 1.12, Conservation of the Black Sea Tursiops truncatus: Bottlenose Dolphin.

ACCOBAMS. (2004). Resolution 2.10, Facilitation of Exchange of Tissue Samples.

ACCOBAMS. (2007a). Resolution 3.9, Guidelines for the Establishment of a System of Tissue Banks within the ACCOBAMS Area and the Ethical Code.

ACCOBAMS. (2007b). Resolution 3.22, Marine Protected Areas for Cetaceans.

ACCOBAMS. (2013a). Resolution 5.14, Live Removals of Bottlenose Dolphins in the Black Sea (Tursiops truncatus).

ACCOBAMS. (2013b). Resolution 5.1, ACCOBAMS long-term strategy 2014–2025.

ACCOBAMS. (2016a). Resolution 6.11, A Strategical Alliance concerning Management and Conservation Measures for the Mediterranean Environment between the Secretariats of ACCOBAMS, GFCM, UNEP/MAP through SPA/RAC, and IUCN-Med, in Collaboration with MedPAN.

ACCOBAMS. (2016b). Resolution 6.21, Species Conservation and Management Plans.

ACCOBAMS. (2019). Resolution 7.17, Global post 2020 biodiversity framework: ACCOBAMS mobilization.

Amalfitano, C. (2018). *General principles of EU law and the protection of fundamental rights*. Cheltenham: Edward Elgar Publishers.

Banaszewska, D. M. (2015). Lex Specialis. In R. Wolfrum (Ed.), *Max Planck Encyclopedia of Public International Law* (online).

Barcelona Convention. (2008). Decision IG. 17/11, "Action Plan for the Conservation of Mediterranean Marine Turtles", the revised implementation timetable of the "Action Plan for the conservation of Bird species listed in Annex II of the Protocol concerning Specially Protected Areas and Biological Diversity in the Mediterranean", and the revised implementation timetable of the "Action Plan concerning species introductions and invasive species in the Mediterranean Sea".

Barcelona Convention. (2016a). Decision IG.22/13, Roadmap for a Comprehensive Coherent Network of Well-Managed Marine Protected Areas (MPAs) to Achieve Aichi Target 11 in the Mediterranean.

Barcelona Convention. (2016b). Decision IG.22/12, Updated Action Plans Concerning "Cetaceans", "Coralligenous and Other Calcareous Bioconcretions", and "Species Introductions and Invasive Species"; Mandate for update of the "Action Plan on Marine and Coastal Birds" and revision of the "Reference List of Marine and Coastal Habitat Types in the Mediterranean".

Barcelona Convention. (2016c). Decision IG.22/18, Cooperation and Partners.

Barcelona Convention. (2017). Decision IG.23/6, Mediterranean quality status report.

Barcelona Convention. (2019). Decision IG.24/07, Strategies and Action Plans under the Protocol concerning Specially Protected Areas and Biological Diversity in the Mediterranean, including the SAP BIO, the Strategy on Monk Seal, and the Action Plans concerning Marine Turtles, Cartilaginous Fishes and Marine Vegetation; Classification of Benthic Marine Habitat Types for the Mediterranean Region, and Reference List of Marine and Coastal Habitat Types in the Mediterranean.

Barral, V. (2012). Sustainable development in international law: Nature and operation of an evolutive legal norm. *The European Journal of International Law, 23*(2), 377–400.

Beyerlin, U., Stoll, P. T., & Wolfrum, R. (Eds.). (2006). *Ensuring compliance with multilateral environmental agreements: A dialogue between practitioners and academia*. Leiden: Martinus Nijhoff Publishers.

Bhat, S. S. (2019). A study of the issue of 'relevant rules' of international law for the purposes of interpretation of treaties under article 31(3)(c) of the Vienna Convention on the law of treaties. *International Community Law Review, 21*(2), 190–219.

Bodansky, D. (2009). Is there an international environmental constitution? *Indiana Journal of Global Legal Studies, 16*(2), 565–584.

Buffard, I., & Zemanek, K. (1998). The "object and purpose" of a treaty: An enigma? *Austrian Review of International & European Law, 3*, 311–343.

Burgiel, S. W. (2017). Invasive Alien Species. In E. Morgera & J. Razzaque (Eds.), *Biodiversity and nature protection law* (pp. 282–294). Cheltenham: Edward Elgar Publishers.

Burns, W. (1998). The agreement on the conservation of cetaceans of the black sea, Mediterranean Sea and contiguous Atlantic area (ACCOBAMS): A regional response to the threats facing cetaceans. *Journal of International Wildlife Law & Policy, 1*(1), 113–133.

CBD. (2006). Decision VIII/24, Protected areas.

CBD. (2008). Decision IX/20, Marine and coastal biodiversity.

CBD. (2010). Decision X/2, The Strategic Plan for Biodiversity 2011–2020 and the Aichi Biodiversity Targets.

CBD. (2014). Decision XII/22, Marine and coastal biodiversity: Ecologically or biologically significant marine areas (EBSAs).

CBD. (2018). Decision XIV/11, Invasive alien species.

CBD. (2020a). *Global Biodiversity Outlook 5*. Montreal: CBD.

CBD. (2020b). CBD/POST2020/PREP/2/1, Update of the zero draft of the post-2020 global biodiversity framework.

CMS. (2020). CMS COP13 in 2020 will kick off big year for biodiversity. Secretariat of the Convention on Migratory Species, https://www.cms.int/en/news/cms-cop13-2020-will-kick-big-year-biodiversity

Cooney, R. (2001). CITES and the CBD: Tensions and synergies. *Review of European, Comparative & International Environmental Law, 10*(3), 259–267.

Cuttelod, A., García, N., Abdul Malak, D., Temple, H., Karariya, V. (2008). The Mediterranean: A biodiversity hotspot under threat. In J. C. Vié, C. Hilton-Taylor, & S. N. Stuart (Eds.), *Wildlife in a changing world: An analysis of the 2008 IUCN Red List of Threatened Species* (pp. 89–101). UICN: Gland.

Delcour, L., & Tulmets, E. (Eds.). (2019). *Policy transfer and norm circulation: Towards an interdisciplinary and comparative approach*. London: Routledge.

ECJ. (2004). Commission v. France, Étang de Berre. Judgment of 7 October 2004, case C-239/03, Rec. CJCE, 2004, I, p. 9328.

FAO. (1995). *Code of conduct for responsible fisheries*. Rome: FAO.

FAO. (2019). *The State of the Mediterranean and Black Sea Fisheries 2018*. Rome: FAO.

Fedtke, J. (2006). Legal transplants. In J. Smits (Ed.), *Elgar encyclopedia of comparative law* (pp. 434–437). Cheltenham: Edward Elgar.

Fitzmaurice, M. (2008). Dynamic (evolutive) interpretation of treaties part I. *The Hague Yearbook of International Law, 21*, 101–153.

Fitzmaurice, M. (2009). Dynamic (evolutive) interpretation of treaties part II. *The Hague Yearbook of International Law, 22*, 3–31.

Friedrich, J. (2008). Legal challenges of nonbinding instruments: The case of the FAO Code of Conduct for Responsible Fisheries. *German Law Journal, 9*(11), 1539–1564.

Futhazar, G. (2015). The diffusion of the Strategic Plan for Biodiversity and its Aichi Biodiversity Targets within the biodiversity cluster: An illustration of current trends in the global governance of biodiversity and ecosystems. *Yearbook of International Environmental Law, 25*, 133–166.

GFCM. (2011a). Report of the Thirty-Fifth Session, Appendix P – Terms of the Reference for a Task Force towards a Possible Revision of the GFCM Agreement, of the Rules of Procedure and the Financial Rules.

GFCM. (2011b). GFCM:XXXV/2011/Inf.8, Performance review of the general fisheries commission for the Mediterranean and Black Sea.

GFCM. (2015). Recommendation 39/2015/3 on the establishment of a set of measures to prevent, deter and eliminate illegal, unreported and unregulated fishing in turbot fisheries in the Black Sea.

GFCM. (2016). GFCM:40/2016/Inf.16, GFCM framework for cooperation and arrangement with non-contracting parties and party organizations.

Haas, P. (1990). *Saving the Mediterranean: The politics of international environmental cooperation*. New York: Columbia University Press.

ICJ. (1980). USA v. Iran, United States Diplomatic and Consular Staff in Tehran. Judgment of 24 May 1980, I.C.J. Reports 1980, 3.
ICJ. (2003). Iran v. USA, Case concerning oil platforms. Judgment of 6 November 2003, I.C.J Reports 2003, 161.
ICJ. (2010). Argentina v. Uruguay, Pulp Mills on the River Uruguay. Judgment of 20 April 2010, ICJ Reports 2010, 14.
ILC. (2006). UNGA A/CN.4/L.682, Fragmentation of international law: Difficulties arising from the diversification and expansion of international law.
IPBES. (2019). *Global assessment report on biodiversity and ecosystem services summary for policy makers*. Bonn: IPBES.
Kiss, A. (1993). Les traités-cadres: une technique juridique caractéristique du droit international de l'environnement. *Annuaire Français de Droit International, 39*, 792–797.
Klabbers, J., Peters, A., & Ulfstein, G. (2011). *The constitutionalization of international law*. Oxford: Oxford University Press.
Klein, D., Carazo, M. P., Meinhard, D., Bulmer, J., Higham, A. (2017). *The Paris Agreement on Climate Change: Analysis and commentary*. Oxford: Oxford University Press.
Kotzé, L. (2019). A global environmental constitution for the Anthropocene? *Transnational Environmental Law, 8*(1), 11–33.
Krämer, L., & Orlando, E. (Eds.). (2018). *Principles of environmental law*. Cheltenham: Edward Elgar Publishers.
Krasner, S. (1983). *International regimes*. London: Cornell University Press.
Kulick, A. (2016). From problem to opportunity? An analytical framework for vagueness and ambiguity in international law. *German Yearbook of International Law, 59*(1), 257–288.
Maes, F., Cliquet, A., du Plessis, W., & McLeod-Kilmurray, H. (Eds.). (2013). *Biodiversity and climate change: Linkages at international, national and local levels*. Cheltenham: Edward Elgar Publishers.
Maljean-Dubois, S. (Ed.). (2017a). *The effectiveness of environmental law*. London: Intersentia.
Maljean-Dubois, S. (Ed.). (2017b). *Circulations de normes et réseaux d'acteurs dans la gouvernance internationale de l'environnement*. Aix-en-Provence: Confluence des Droits.
MAP. (2003). *Strategic Action Programme for the Conservation of Biological Diversity (SAP BIO) in the Mediterranean Region*. Tunis: RAC–SPA.
MAP. (2008). UNEP(DEPI)/MED IG.17/Inf.17, Guidelines for controlling the vectors of introduction into the Mediterranean of non-indigenous species and invasive marine species.
MAP. (2012). *State of the Mediterranean marine and coastal environment*. Athens: MAP.
MAP. (2019a). UNEP/MED WG.468/Inf.16, Draft 2019 State of the Environment and Development Report.
MAP. (2019b). UNEP/MED WG.468/Inf 11, Evaluation of the implementation of the Strategic Action Programme for the Conservation of Biological Diversity in the Mediterranean Region (SAP BIO) and orientations for the elaboration of a post-2020 SAP BIO, as reviewed by the Fourteenth Meeting of SPA/BD Thematic Focal Points.
Matz-Lück, N. (2009). Framework conventions as regulatory tool. *Goettingen Journal of International Law, 1*(3), 439–458.
McLachlan, C. (2005). The principle of systemic integration and Article 31(3)(c) of the Vienna Convention. *International & Comparative Law Quarterly, 54*(2), 279–320.
Miller, J. M. (2003). A typology of legal transplants: Using sociology, legal history and Argentine examples to explain the transplant process. *The American Journal of Comparative Law, 51*(4), 839–885.
Morgera, E. (2015). Global environmental law and comparative legal methods. *Review of European, Comparative & International Environmental Law, 24*(3), 254–263.
Ovodenko, A., & Keohane, R. (2012). Institutional diffusion in international environmental affairs. *International Affairs, 88*(3), 523–541.

Parks, L., & Morgera, E. (2015). The need for an interdisciplinary approach to norm diffusion: The case of fair and equitable benefit-sharing. *Review of European, Comparative & International Environmental Law, 24*(3), 353–367.

Pasquet, L. (2018). De-fragmentation techniques. In H. Ruiz Fabri (Ed.), *Max Planck Encyclopedia of International Procedural Law* (online).

Peñas Lado, E. (2016). *The common fisheries policy: The quest for sustainability*. Oxford: Wiley-Blackwell.

Perelman, C., & Vander Elst, R. (Eds.). (1984). *Les notions à contenu variable en droit*. Brussels: Bruylant.

Peters, A. (2017). (2019). Constitutionalization. In d'Aspremont, J., Singh, S. (Eds.), Concepts for international law: Contributions to disciplinary thought. Cheltenham: Edward Elgar Publishers, 141-153.

Riley, S. (2010). A weed by any other name: Would the rose smell as sweet if it were a threat to biodiversity? *Georgetown International Environmental Law Review, 22*(1), 157–183.

Rogalla von Bieberstein, K., Sattout, E., Christensen, M., Pisupati, B., Burgess, N., Harrison, J., Geldmann, J. (2018). Improving collaboration in the implementation of global biodiversity conventions. *Conservation Biology, 33*(4), 821–831.

Ruiz Fabri, H., & Gradoni, L. (Eds.). (2009). *La circulation des concepts juridiques Le droit international de l'environnement entre mondialisation et fragmentation*. Paris: LGDJ.

Scovazzi, T. (1999). Regional cooperation in the field of the environment. In T. Scovazzi (Ed.), *Marine specially protected areas: The general aspects and the Mediterranean regional system* (pp. 81–99). The Hague: Kluwer Law International.

Tavares de Pinho, A. (1997). La Réforme de la Commission Générale des Pêches pour la Méditerranée. *Annuaire de Droit de la Mer, 2*, 65–91.

Teubner, G. (2012). *Constitutional fragments: Societal constitutionalism and globalization*. Oxford: Oxford University Press.

Twining, W. (2013). Diffusion of law: A global perspective. *The Journal of Legal Pluralism and Unofficial Law, 36*(49), 1–45.

UNEP. (1991). UNEP/BIODIV/N4-INC2/2, Second revised draft convention on biological diversity.

WTO. (1998). United States – Import Prohibition of Certain Shrimp and Shrimp Products. Panel report of 15 May 1998, WT/DS58/R, Appellate Body Report of 12 October 1998, WT/DS58/AB/R.

Young, M. (2012). *Regime interaction in international law: Facing fragmentation*. Cambridge: Cambridge University Press.